Lecture Notes
in Control and Information Sciences   451

T0213899

For further volumes:
www.springer.com/series/642

Lecture Notes
in Control and Information Sciences

Editors

Professor Dr.-Ing. Manfred Thoma
Institut für Regelungstechnik, Universität Hannover, Appelstr. 11, 30167 Hannover,
Germany
E-mail: thoma@irt.uni-hannover.de

Prof. Dr. Frank Allgöwer
Institute for Systems Theory and Automatic Control, University of Stuttgart,
70550 Stuttgart, Germany
E-mail: allgower@ist.uni-stuttgart.de

Professor Dr. Manfred Morari
ETH/ETL I 29, Physikstr. 3, 8092 Zürich, Switzerland
E-mail: morari@aut.ee.ethz.ch

Series Advisory Board

P. Fleming
University of Sheffield, Sheffield, UK

P. Kokotovic
University of California, Santa Barbara, CA, USA

A.B. Kurzhanski
Moscow State University, Moscow, Russia

H. Kwakernaak
University of Twente, Enschede, The Netherlands

A. Rantzer
Lund Institute of Technology, Lund, Sweden

J.N. Tsitsiklis
MIT, Cambridge, MA, USA

Hoai-Nam Nguyen

Constrained Control of Uncertain, Time-Varying, Discrete-Time Systems

An Interpolation-Based Approach

 Springer

Hoai-Nam Nguyen
Technion—Israel Institute of Technology
Haifa, Israel

ISSN 0170-8643   ISSN 1610-7411 (electronic)
Lecture Notes in Control and Information Sciences
ISBN 978-3-319-02826-2   ISBN 978-3-319-02827-9 (eBook)
DOI 10.1007/978-3-319-02827-9
Springer Cham Heidelberg New York Dordrecht London

Library of Congress Control Number: 2013957060

Printed on acid-free paper

Springer is part of Springer Science+Business Media (www.springer.com)

To Linh and Tung

Preface

A fundamental problem in automatic control is the control of uncertain and/or time-varying plants with input and state or output constraints. Most elegantly, and theoretically most satisfying, the problem is solved by optimal control which, however, rarely gives a feedback solution, and oftentimes only a numerical solutions.

Therefore, in practice, the problem has seen many ad-hoc solutions, such as *override control, anti-windup*. Another solution, that has become popular during the last decades is *Model Predictive Control* (MPC) where an optimal control problem is solved at each sampling instant, and the element of the control vector meant for the nearest sampling interval is applied. In spite of the increased computational power of control computers, MPC is at present mainly suitable for low-order, nominally linear systems. The robust version of MPC is conservative and computationally complicated, while the *explicit* version of MPC that gives a piecewise affine state feedback solution involves a very complicated division of the state space into polyhedral cells.

In this book a novel and computationally cheap solution is presented for uncertain and/or time-varying linear discrete-time systems with polytopic bounded control and state (or output) vectors, with bounded disturbances. The approach is based on the interpolation between a stabilizing, outer low-gain controller that respects the control and state constraints, and an inner, high-gain controller, designed by any method that has its robustly positively invariant set satisfying the constraints. A simple Lyapunov function is used for the proof of closed loop stability.

In contrast to MPC, the new interpolating controller is not necessarily employing an optimization criterion inspired by performance. In its explicit form, the cell partitioning is considerable simpler that the MPC counterpart. For the implicit version, the on-line computational demand can be restricted to the solution of at most two linear programming problems or one quadratic programming problem or one semi-definite programming problem.

Several simulation examples are given, including uncertain linear systems with output feedback and disturbances. Some examples are compared with MPC. It is believed that the new controller might see wide-spread use in industry, including the automotive industry, also for the control of fast, high-order systems with constraints.

Haifa, Israel   Hoai-Nam Nguyen

Acknowledgements

This book is a revised version of my PhD thesis from the automatic control department at Ecole Supérieure d'Electricité, Supélec, Gif-sur-Yvette, France, written in the spring of 2012. My thesis advisors were Sorin Olaru and Per-Olof Gutman. Almost all the material presented in this book is based on joint work with them.

I would like first to thank Sorin Olaru, for his enthusiastic support, and for creating and maintaining a creative environment for research and studies. To him and his close collaborator, Professor Morten Hovd, I am grateful for their help which helped me to take the first steps into becoming a scientist.

Big thanks and a big hug goes to Per-Olof Gutman, who always willingly shared his deep and vast knowledge with me. Thanks to him, I believe that I learned to think the right way about some problems in control, which I would not be able to approach appropriately otherwise.

In the automatic control department at Supelec, I would like to thank especially to Professor Patrick Boucher for his support. I would also like to thank to all the people I have met in the department, particularly to Professor Didier Dumur, Professor Gilles Duc, Cristina Stoica, Pedro Rodriguez, Hieu, Alaa, Florin, Ionela, Nikola.

Finally, the biggest thanks goes to my wife Linh and my son Tung, for their never ending support. I dedicate this book to them, in love and gratitude.

Contents

Part II   Interpolating Control

Notation[1]

Sets

$\mathbb{R}$	Set of real number
$\mathbb{R}_+$	Set of nonnegative real number
$\mathbb{R}^n$	Set of real vectors with n elements
$\mathbb{R}^{n \times m}$	Set of real matrices with n rows and m columns

Algebraic Operators

A^T	Transpose of matrix A
A^{-1}	Inverse of matrix A
$A \succ (\succeq)0$	Positive (semi)definite matrix
$A \prec (\preceq)0$	Negative (semi)definite matrix

Set Operators

$P_1 \cap P_2$	Set intersection
$P_1 \oplus P_2$	Minkowski sum
$P_1 \ominus P_2$	Pontryagin difference
$P_1 \subseteq P_2$	P_1 is a subset of P_2
$P_1 \subset P_2$	P_1 is a strict subset of P_2
$P_1 \supseteq P_2$	P_1 is a superset of P_2
$P_1 \supset P_2$	P_1 is a strict superset of P_2
∂P	The boundary of P
$\text{Int}(P)$	The interior of P
$\text{Proj}_x(P)$	The orthogonal projection of the set P onto the x space

Others

$\mathbf{I}$	Identity matrix of appropriate dimension
$\mathbf{1}$	Matrix of ones of appropriate dimension
$\mathbf{0}$	Matrix of zeros of appropriate dimension

---

[1] The conventions and the notations used in the book are classical for the control literature. A short description is provided in the following.

Acronyms

LMI	Linear Matrix Inequality
LP	Linear Programming
QP	Quadratic Programming
LQR	Linear Quadratic Regulator
LTI	Linear Time Invariant
LPV	Linear Parameter Varying
PWA	PieceWise Affine

Chapter 1
Introduction

Constraints are encountered practically in all real-world control problems. The presence of constraints leads to theoretical and computational challenges. From the conceptual point of view, constraints can have different nature. Basically, there are two types of constraints imposed by physical limitation and/or performance desiderata.

Physical constraints are due to the physical limitations of the mechanical, electrical, biological, etc controlled systems. The input and output constraints must be fulfilled to avoid over-exploitation or damage. In addition, the constraint violation may lead to degraded performance, oscillations or even instability.

Performance constraints are introduced by the designer for guaranteeing performance requirements, e.g. transient time, transient overshoot, etc.

The constrained control problem becomes even more challenging in the presence of model uncertainties which is unavoidable in practice [1, 117]. It is generally accepted that a key reason of using feedback is to diminish the effects of uncertainty which may appear in different forms as disturbances or as other inadequacies in the models used to design the feedback law. Model uncertainty and robustness have been a central theme in the development of the field of automatic control [8].

A straightforward way to stabilize a system with input constraints is to perform the control design disregarding the constraints, then an adaptation of the control law is considered with respect to input saturation. Such an approach is called *anti-windup* [73, 123, 124, 132].

Over the last decades, the research on constrained control topics has developed to the degree that constraints can be taken into account during the synthesis phase. By its principle, model predictive control (MPC) approach shows its importance on dealing with constraints [2, 30, 34, 47, 48, 88, 92, 107]. In MPC, a sequence of predicted optimal control values over a finite prediction horizon is computed for optimizing the performance of the controlled system, expressed in terms of a cost function. Then only the first element of the optimal sequence is actually applied to the system and the entire optimization is repeated at the next time instant with the new state measurement [4, 88, 92].

In MPC, with a linear model, polyhedral constraints, and a quadratic cost, the resulting optimization problem is a quadratic programming (QP) problem [37, 104].

H.-N. Nguyen, *Constrained Control of Uncertain, Time-Varying, Discrete-Time Systems*,
Lecture Notes in Control and Information Sciences 451,
DOI 10.1007/978-3-319-02827-9_1,

Solving the QP problem can be computationally costly, specially when the prediction horizon is large, and this has traditionally limited MPC to applications with relatively low complexity/sampling interval ratio [3].

In the last decade, attempts have been made to use predictive control in fast processes. In [20, 100, 101, 127] it was shown that the constrained linear MPC is equivalent to a multi-parametric optimization problem, where the state plays the role of a vector of parameters. The solution is a piecewise affine function of the state over a polyhedral partition of the state space, and the computational effort of MPC is moved off-line. This control law is called *explicit MPC*. However, explicit MPC also has disadvantages. Obtaining the explicit optimal MPC solution requires to solve an off-line parametric optimization problem, which is NP-hard. Although the problem is tractable and practically solvable for several interesting control applications, the off-line computational effort grows *exponentially* fast as the problem size increases [61, 62, 77–79].

In [131], it was shown that the on-line computation is preferable for high dimensional systems where significant reduction of the computational complexity can be achieved by exploiting the particular structure of the optimization problem as well as by early stopping and warm-starting from a solution obtained at the previous time-step. The same reference mentioned that for models of more than five dimensions the explicit solution might be impractical. It is worth mentioning that approximate explicit solutions have been investigated to go beyond this ad-hoc limitation [18, 60, 114].

Note that as its name says, most traditional implicit and explicit MPC approaches are based on mathematical models which invariably present a mismatch with respect to the physical systems. The robust MPC is meant to address both model uncertainty and disturbances. However, the robust MPC presents great conservativeness and/or on-line computational burden [21, 35, 74, 81, 84].

The use of interpolation in constrained control in order to avoid very complex control design procedures is well known in the literature. There is a long line of developments on these topics generally closely related to MPC, see for example [10, 102, 108–110], where interpolation between input sequences, state trajectories, different feedback gains with associated invariant sets can be found.

The vertex control law can be considered also as an interpolation approach based on the admissible control values, assumed to be available for the vertices of a polyhedral positively invariant set C_N in the state space [23, 53]. A weakness of vertex control is that the full control range is exploited only on the border of C_N, and hence the time to regulate the plant to the origin is much longer than e.g. by time-optimal control. A way to overcome this shortcoming is to switch to another, more aggressive, local controller near the origin in the state space, e.g. a state feedback controller $u = Kx$, when the state reaches the Maximal Admissible Set (MAS) of the local controller. The disadvantage of such a switching-based solution is that the control action becomes non-smooth [94].

The aim of this book is to propose an alternative to MPC. The book gives a comprehensive development of the novel Interpolating Control (IC) of the regulation problem for linear, time-invariant, discrete-time uncertain dynamical systems with polyhedral state space, and polyhedral control constraints, with and without disturbances, under state- or output feedback. For output feedback a non-minimal state-space representation is used with old inputs and outputs as state variables.

The book is structured in three parts. The first part includes background material and the theoretical foundation of the work. Beyond a briefly review of the area, some new results on estimating the domain of attraction are provided.

The second part contains the main body of the book, with three chapters on interpolating control that addresses nominal state feedback, robust state feedback and output feedback, respectively. The IC is given in both its implicit form, where at most two LP-problems or one QP-problem or one semi-definite problem are solved on-line at each sampling instant to yield the value of the control variable, and in its explicit form where the control law is shown to be piecewise affine in the state, whereby the state space is partitioned in polyhedral cells and whereby at each sampling interval it has to be determined to which cell the measured state belongs. The interpolation in IC is performed between constraint-aware low-gain feedback strategies in order to respect the constraints, and a user-chosen performance control law in its MAS surrounding the origin.

Thus, IC is composed by an *outer* control law (the interpolated control) near the boundaries of the allowed state set whose purpose is to make the state enter the MAS rapidly (but not necessarily time-optimally) without violating any constraints, and the *inner* user-chosen control law in its MAS whose purpose is performance according to the user's choice.

Novel proofs of recursive feasibility and asymptotic stability of the Vertex Control law, and of the Interpolating Control law are given. Algorithms for Implicit and Explicit IC are presented in such a way that the reader may easily realize them.

Each chapter includes illustrative examples, and comparisons with MPC. It is demonstrated that the computation complexity of IC is considerably lower than that of MPC, in particular for high-order systems, and systems with uncertainty, although the performance is similar except in those cases when an MPC solution cannot be found.

In the last part of the book two high order examples as well as a benchmark problem of robust MPC are reported in order to illustrate some practical aspects of the proposed methods.

Part I
Background

Chapter 2
Set Theoretic Methods in Control

2.1 Set Terminology

For completeness, some standard definitions of set terminology will be introduced. For a detailed reference, the reader is referred to the book [72].

Definition 2.1 (Closed set) A set $S \subseteq \mathbb{R}^n$ is *closed* if it contains its own boundary. In other words, any point outside S has a neighborhood disjoint from S.

Definition 2.2 (Closure of a set) The *closure* of a set $S \subseteq \mathbb{R}^n$ is the intersection of all *closed sets* containing S.

Definition 2.3 (Bounded set) A set $S \subset \mathbb{R}^n$ is *bounded* if it is contained in some ball $B_R = \{x \in \mathbb{R}^n : \|x\|_2 \leq \varepsilon\}$ of finite radius $\varepsilon > 0$.

Definition 2.4 (Compact set) A set $S \subset \mathbb{R}^n$ is *compact* if it is closed and bounded.

Definition 2.5 (Support function) The support function of a set $S \subset \mathbb{R}^n$, evaluated at $z \in \mathbb{R}^n$ is defined as

$$\phi_S(z) = \sup_{x \in S} \{z^T x\}$$

2.2 Convex Sets

2.2.1 Basic Definitions

The fact that convexity is a more important property than linearity has been recognized in several domains, the optimization theory being maybe the best example [31, 106]. We provide in this section a series of definitions which will be useful in the sequel.

H.-N. Nguyen, *Constrained Control of Uncertain, Time-Varying, Discrete-Time Systems*,
Lecture Notes in Control and Information Sciences 451,
DOI 10.1007/978-3-319-02827-9_2,
© Springer International Publishing Switzerland 2014

Definition 2.6 (Convex set) A set $S \subset \mathbb{R}^n$ is *convex* if it holds that, $\forall x_1 \in S$ and $\forall x_2 \in S$,

$$\alpha x_1 + (1 - \alpha)x_2 \in S, \quad \forall \alpha \in [0, 1]$$

The point

$$x = \alpha x_1 + (1 - \alpha)x_2$$

where $0 \leq \alpha \leq 1$ is called a *convex combination* of the pair $\{x_1, x_2\}$. The set of all such points is the line segment connecting x_1 and x_2. Obviously, a set S is convex if a segment between any two points in S lies in S.

Definition 2.7 (Convex function) A function $f : S \rightarrow \mathbb{R}$ with a convex set $S \subseteq \mathbb{R}^n$ is *convex* if and only if, $\forall x_1 \in S$, $\forall x_2 \in S$ and $\forall \alpha \in [0, 1]$,

$$f\big(\alpha x_1 + (1 - \alpha)x_2\big) \leq \alpha f(x_1) + (1 - \alpha)f(x_2)$$

Definition 2.8 (C-set) A set $S \subset \mathbb{R}^n$ is a *C*-set if it is a *convex* and *compact* set, containing the origin in its interior.

Definition 2.9 (Convex hull) The *convex hull* of a set $S \subset \mathbb{R}^n$ is the smallest convex set containing S.

It is well known [133] that for any finite set $S = \{s_1, s_2, \ldots, s_r\}$, where $s_i \in \mathbb{R}^n$, $i = 1, 2, \ldots, r$, the convex hull of S is given by

$$\text{Convex Hull}(S) = \left\{ s \in \mathbb{R}^n : s = \sum_{i=1}^{r} \alpha_i s_i : \forall s_i \in S \right\}$$

where $\sum_{i=1}^{r} \alpha_i = 1$ and $\alpha_i \geq 0$, $i = 1, 2, \ldots, r$.

2.2.2 Ellipsoidal Set

Ellipsoidal sets or ellipsoids are one of the famous classes of convex sets. Ellipsoids represent a large category used in the study of dynamical systems due to their simple numerical representation [32, 75]. Next we provide a formal definition for ellipsoids and a few properties.

Definition 2.10 (Ellipsoidal set) An ellipsoid $E(P, x_0) \subset \mathbb{R}^n$ with center x_0 and shape matrix P is a set of the form,

$$E(P, x_0) = \left\{ x \in \mathbb{R}^n : (x - x_0)^T P^{-1} (x - x_0) \leq 1 \right\} \tag{2.1}$$

where $P \in \mathbb{R}^{n \times n}$ is a positive definite matrix.

If $x_0 = 0$ then it is possible to write,

$$E(P) = \left\{ x \in \mathbb{R}^n : x^T P^{-1} x \leq 1 \right\} \tag{2.2}$$

Define $Q = \sqrt{P}$ as the Cholesky factor of matrix P, which satisfies

$$Q^T Q = Q Q^T = P$$

With matrix Q, it is possible to show an alternative dual representation of ellipsoid (2.1)

$$D(Q, x_0) = \left\{ x \in \mathbb{R}^n : x = x_0 + Qz \right\}$$

where $z \in \mathbb{R}^n$ such that $z^T z \leq 1$.

Ellipsoids are probably the most commonly used in the control field since they are associated with powerful tools such as Linear Matrix Inequalities (LMI) [32, 112]. When using ellipsoids, almost all the control optimization problems can be reduced to the optimization of a linear function under LMI constraints. This optimization problem is convex and is by now a powerful design tool in many control applications.

A linear matrix inequality is a condition of the type [32, 112],

$$F(x) \succeq 0$$

where $x \in \mathbb{R}^n$ is a vector variable and

$$F(x) = F_0 + \sum_{i=1}^{n} F_i x_i$$

with symmetric matrices $F_i \in \mathbb{R}^{m \times m}$.

LMIs can either be understood as feasibility conditions or constraints for optimization problems. Optimization of a linear function over LMI constraints is called semi-definite programming, which is considered as an extension of linear programming. Nowadays, a major benefit in using LMIs is that for solving an LMI problem, several polynomial time algorithms were developed and implemented in free available software packages, such as LMI Lab [43], YALMIP [87], CVX [49], SEDUMI [121], etc.

The Schur complements are a very useful tool for manipulating matrix inequalities. The Schur complements state that the nonlinear conditions of the special forms,

$$\begin{cases} P(x) \succ 0 \\ P(x) - Q(x)^T R(x)^{-1} Q(x) \succ 0 \end{cases} \tag{2.3}$$

or

$$\begin{cases} R(x) \succ 0 \\ R(x) - Q(x) P(x)^{-1} Q(x)^T \succ 0 \end{cases} \tag{2.4}$$

can be equivalently written in the LMI form,

$$\begin{bmatrix} P(x) & Q(x)^T \\ Q(x) & R(x) \end{bmatrix} \succ 0 \tag{2.5}$$

The Schur complements allows one to convert certain nonlinear matrix inequalities into a higher dimensional LMI. For example, it is well known [75] that the support function of the ellipsoid $E(P)$, evaluated at the vector $f \in \mathbb{R}^n$ is,

$$\phi_{E(P)}(z) = \sqrt{f^T P f} \tag{2.6}$$

then clearly, $E(P)$ is a subset of the polyhedral set[1] $\mathscr{P}(f, 1)$, where

$$\mathscr{P}(f, 1) = \left\{ x \in \mathbb{R}^n : \left| f^T x \right| \le 1 \right\}$$

if and only if

$$f^T P f \le 1$$

or by using the Schur complements this condition can be rewritten as [32, 55],

$$\begin{bmatrix} 1 & f^T P \\ Pf & P \end{bmatrix} \succeq 0 \tag{2.7}$$

An ellipsoid $E(P, x_0) \subset \mathbb{R}^n$ is uniquely defined by its matrix P and by its center x_0. Since matrix P is symmetric, the complexity of the representation (2.1) is

$$\frac{n(n + 1)}{2} + n = \frac{n(n + 3)}{2}$$

The main drawback of ellipsoids is however that having a fixed and symmetrical structure they may be too conservative and this conservativeness is increased by the related operations. It is well known [75] that[2]

- The convex hull of of a set of ellipsoids, in general, is not an ellipsoid.
- The sum of two ellipsoids is not, in general, an ellipsoid.
- The difference of two ellipsoids is not, in general, an ellipsoid.
- The intersection of two ellipsoids is not, in general, an ellipsoid.

2.2.3 Polyhedral Set

Polyhedral sets provide a useful geometrical representation for linear constraints that appear in diverse fields such as control and optimization. In a convex setting,

---

[1] A rigorous definition of polyhedral sets will be given in Sect. 2.2.3.

[2] The reader is referred to [75] for the definitions of operations with ellipsoids.

they provide a good compromise between complexity and flexibility. Due to their linear and convex nature, the basic set operations are relatively easy to implement [76, 129]. Principally, this is related to their dual (half-spaces/vertices) representation [33, 93] which allows choosing which formulation is best suited for a particular problem. This section is started by recalling some theoretical concepts.

Definition 2.11 (Hyperplane) A hyperplane $H(f, g)$ is a set of the form,

$$H(f, g) = \left\{ x \in \mathbb{R}^n : f^T x = g \right\} \tag{2.8}$$

where $f \in \mathbb{R}^n$, $g \in \mathbb{R}$.

Definition 2.12 (Half-space) A closed half-space $\mathscr{H}(f, g)$ is a set of the form,

$$\mathscr{H}(f, g) = \left\{ x \in \mathbb{R}^n : f^T x \le g \right\} \tag{2.9}$$

where $f \in \mathbb{R}^n$, $g \in \mathbb{R}$.

Definition 2.13 (Polyhedral set)   A convex polyhedral set $P(F, g)$ is a set of the form,

$$P(F, g) = \left\{ x \in \mathbb{R}^n : F_i^T x \le g_i, \ i = 1, 2, \dots, n_1 \right\} \tag{2.10}$$

where $F_i^T \in \mathbb{R}^n$ denotes the i-th row of the matrix $F \in \mathbb{R}^{n_1 \times n}$ and g_i is the i-th component of the column vector $g \in \mathbb{R}^{n_1}$.

A polyhedral set contains the origin if and only if $g \ge 0$, and includes the origin in its interior if and only if $g > 0$.

Definition 2.14 (Polytope) A polytope is a *bounded* polyhedral set.

Definition 2.15 (Dimension of polytope) A polytope $P(F, g) \subset \mathbb{R}^n$ is of dimension $d \le n$, if there exists a d-dimension ball with radius $\varepsilon > 0$ contained in $P(F, g)$ and there exists no $(d + 1)$-dimension ball with radius $\varepsilon > 0$ contained in $P(F, g)$. A polytope is full dimensional if $d = n$.

Definition 2.16 (Face, facet, vertex, edge) An $(n - 1)$-dimensional face F_{ai} of polytope $P(F, g) \subset \mathbb{R}^n$ is defined as,

$$F_{ai} = \left\{ x \in P : F_i^T x = g_i \right\} \tag{2.11}$$

and can be interpreted as the intersection between P and a *non-redundant supporting hyperplane*

$$F_{ai} = P \cap \left\{ x \in \mathbb{R}^n : F_i^T x = g_i \right\} \tag{2.12}$$

The non-empty intersection of two faces of dimension $(n - r)$ with $r = 0, 1, \dots,$ $n - 1$ leads to the description of $(n - r - 1)$-dimensional face. The faces of $P(F, g)$ with dimension 0, 1 and $(n - 1)$ are called *vertices*, *edges* and *facets*, respectively.

Fig. 2.1 Exemplification for
the equivalent of half-space
and vertex representations of
polytopes

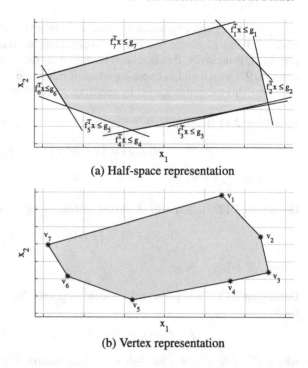

(a) Half-space representation

(b) Vertex representation

One of the fundamental properties of polytope is that it can be presented in half-space representation as in Definition 2.13 or in vertex representation as

$$P(V) = \left\{ x \in \mathbb{R}^n : x = \sum_{i=1}^{r} \alpha_i v_i \right\}$$

where $v_i \in \mathbb{R}^n$ is the i-column of matrix $V \in \mathbb{R}^{n \times r}$, $\sum_{i=1}^{r} \alpha_i = 1$ and $\alpha_i \geq 0$, $i = 1, 2, \ldots, r$, see Fig. 2.1.

Note that the transformation from half-space (vertex) representation to vertex (half-space) representation may be time-consuming with several well-known algorithms: Fourier-Motzkin elimination [39], CDD [41], Equality Set Projection [64].

Recall that the expression $x = \sum_{i=1}^{r} \alpha_i v_i$ with a given set of vectors $\{v_1, v_2, \ldots, v_r\}$ and

$$\sum_{i=1}^{r} \alpha_i = 1, \quad \alpha_i \geq 0$$

is called *the convex hull* of the set $\{v_1, v_2, \ldots, v_r\}$ and will be denoted as

$$x = \text{Conv}\{v_1, v_2, \ldots, v_r\}$$

Definition 2.17 (Simplex) A simplex $S \subset \mathbb{R}^n$ is an n-dimensional polytope, which is the convex hull of $n + 1$ vertices.

For example, a $2D$-simplex is a triangle, a $3D$-simplex is a tetrahedron, and a $4D$-simplex is a pentachoron.

Definition 2.18 (Redundant half-space) For a given polytope $P(F, g)$, the polyhedral set $P(\overline{F}, \overline{g})$ is defined by removing the i-th half-space F_i^T from matrix F and the corresponding component g_i from vector g. The facet (F_i^T, g_i) is *redundant* if and only if

$$\overline{g}_i < g_i \tag{2.13}$$

where

$$\overline{g}_i = \max_x \{F_i^T x\} \quad \text{s.t. } x \in P(\overline{F}, \overline{g})$$

Definition 2.19 (Redundant vertex) For a given polytope $P(V)$, the polyhedral set $P(\overline{V})$ is defined by removing the i-th vertex v_i from the matrix V. The vertex v_i is *redundant* if and only if

$$p_i < 1 \tag{2.14}$$

where

$$p_i = \min_p \{1^T p\} \quad \text{s.t. } \begin{cases} \overline{V}p = v_i, \\ p \geq 0 \end{cases}$$

Definition 2.20 (Minimal representation) A half-space or vertex representation of polytope P is *minimal* if and only if the removal of any facet or any vertex would change P, i.e. there are no redundant facets or redundant vertices.

Clearly, a minimal representation of a polytope can be achieved by removing from the half-space (vertex) representation all the redundant facets (vertices).

Definition 2.21 (Normalized representation) A polytope

$$P(F, g) = \{x \in \mathbb{R}^n : F_i^T x \leq g_i, \ i = 1, 2, \ldots, n_1\}$$

is in a normalized representation if it has the following property

$$F_i^T F_i = 1, \quad \forall i = 1, 2, \ldots, n_1$$

A normalized full dimensional polytope has a *unique* minimal representation. This fact is very useful in practice, since normalized full dimensional polytopes in minimal representation allow us to avoid any ambiguity when comparing them.

Next, some basic operations on polytopes will be briefly reviewed. Note that although the focus lies on polytopes, most of the operations described here are directly or with minor changes applicable to polyhedral sets. Additional details on polytope computation can be found in [42, 51, 133].

Fig. 2.2 Minkowski sum and
Pontryagin difference of
polytopes

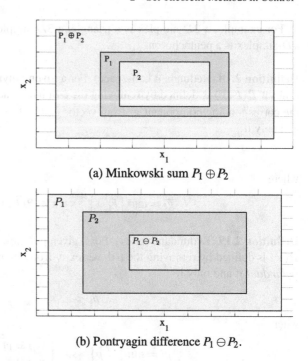

(a) Minkowski sum $P_1 \oplus P_2$

(b) Pontryagin difference $P_1 \ominus P_2$.

Definition 2.22 (Intersection) The intersection of two polytopes $P_1 \subset \mathbb{R}^n$, $P_2 \subset \mathbb{R}^n$ is a polytope,

$$P_1 \cap P_2 = \left\{ x \in \mathbb{R}^n : x \in P_1, x \in P_2 \right\}$$

Definition 2.23 (Minkowski sum) The Minkowski sum of two polytopes $P_1 \subset \mathbb{R}^n$, $P_2 \subset \mathbb{R}^n$ is a polytope, see Fig. 2.2(a),

$$P_1 \oplus P_2 = \{x_1 + x_2 : x_1 \in P_1, x_2 \in P_2\}$$

It is well known [133] that if P_1 and P_2 are given in vertex representation, i.e.

$$P_1 = \text{Conv}\{v_{11}, v_{12}, \ldots, v_{1p}\},$$
$$P_2 = \text{Conv}\{v_{21}, v_{22}, \ldots, v_{2q}\}$$

then their Minkowski sum can be computed as,

$$P_1 \oplus P_2 = \text{Conv}\{v_{1i} + v_{2j}\}, \quad \forall i = 1, 2, \ldots, p, \ \forall j = 1, 2, \ldots, q$$

Definition 2.24 (Pontryagin difference) The Pontryagin difference of two polytopes $P_1 \subset \mathbb{R}^n$, $P_2 \subset \mathbb{R}^n$ is the polytope, see Fig. 2.2(b),

$$P_1 \ominus P_2 = \{x_1 \in P_1 : x_1 + x_2 \in P_1, \forall x_2 \in P_2\}$$

Fig. 2.3 Projection of a
2-dimensional polytope P
onto a line x_1

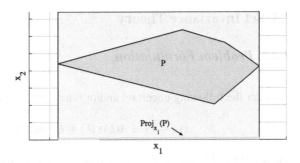

Note that the Pontryagin difference is not the complement of the Minkowski sum. For two polytopes P_1 and P_2, it holds only that $(P_1 \ominus P_2) \oplus P_2 \subseteq P_1$.

Definition 2.25 (Projection) Given a polytope $P \subset \mathbb{R}^{n_1+n_2}$, the orthogonal projection of P onto the x_1-space $\mathbb{R}^{n_1}$ is defined as, see Fig. 2.3,

$$\text{Proj}_{x_1}(P) = \left\{ x_1 \in \mathbb{R}^{n_1} : \exists x_2 \in \mathbb{R}^{n_2} \text{ such that } \begin{bmatrix} x_1^T & x_2^T \end{bmatrix}^T \in P \right\}$$

It is well known [133] that the Minkowski sum operation on polytopes in their half-space representation is complexity-wise equivalent to a projection. Current projection methods for polytopes that can operate in general dimensions can be grouped into four classes: Fourier elimination [66], block elimination [12], vertex based approaches and wrapping-based techniques [64].

Clearly, the complexity of the representation of polytopes is not a function of the space dimension only, but it may be arbitrarily big. For the half-space (vertex) representation, the complexity of the polytopes is a linear function of the number of rows of the matrix F (the number of columns of the matrix V). As far as the complexity issue concerns, it is worth to be mentioned that none of these representations can be regarded as more convenient. Apparently, one can define an arbitrary polytope with relatively few vertices, however this may nevertheless have a surprisingly large number of facets. This happens, for example when some vertices contribute to many facets. And equally, one can define an arbitrary polytope with relatively few facets, however this may have relatively many more vertices. This happens, for example when some facets have many vertices [42].

The main advantage of the polytopes is their flexibility. It is well known [33] that any convex body can be approximated arbitrarily close by a polytope. Particularly, for a given bounded, convex and closed set S and for a given scalar $0 < \varepsilon < 1$, then there exists a polytope P such that,

$$(1-\varepsilon)S \subseteq P \subseteq S$$

for an inner ε-approximation of the set S and

$$S \subseteq P \subseteq (1+\varepsilon)S$$

for an outer ε-approximation of the set S.

2.3 Set Invariance Theory

2.3.1 Problem Formulation

Consider the following uncertain and/or time-varying linear discrete-time system,

$$x(k+1) = A(k)x(k) + B(k)u(k) + Dw(k) \qquad (2.15)$$

where $x(k) \in \mathbb{R}^n$, $u(k) \in \mathbb{R}^m$, $w(k) \in \mathbb{R}^d$ are, respectively the state, input and disturbance vectors. The matrices $A(k) \in \mathbb{R}^{n \times n}$, $B(k) \in \mathbb{R}^{n \times m}$, $D \in \mathbb{R}^{n \times d}$. $A(k)$ and $B(k)$ satisfy,

$$\begin{cases} A(k) = \sum_{i=1}^{q} \alpha_i(k) A_i, \qquad B(k) = \sum_{i=1}^{q} \alpha_i(k) B_i \\ \sum_{i=1}^{q} \alpha_i(k) = 1, \quad \alpha_i(k) \geq 0 \end{cases} \qquad (2.16)$$

where the matrices A_i, B_i, $i = 1, 2, \ldots, q$ are the extreme realizations of $A(k)$ and $B(k)$.

Theorem 2.1 $A(k)$, $B(k)$ given as,

$$\begin{cases} A(k) = \sum_{i=1}^{q_1} \alpha_i(k) A_i, \qquad B(k) = \sum_{j=1}^{q_2} \beta_j(k) B_j, \\ \sum_{i=1}^{q_1} \alpha_i(k) = 1, \quad \alpha_i(k) \geq 0, \ \forall i = 1, 2, \ldots, q_1, \\ \sum_{j=1}^{q_2} \beta_j(k) = 1, \quad \beta_j(k) \geq 0, \ \forall j = 1, 2, \ldots, q_2 \end{cases} \qquad (2.17)$$

can be transformed into the form of (2.16).

Proof For simplicity, the case $D = 0$ is considered. One has,

$$x(k+1) = \sum_{i=1}^{q_1} \alpha_i(k) A_i x(k) + \sum_{j=1}^{q_2} \beta_j(k) B_j u(k)$$

$$= \sum_{i=1}^{q_1} \alpha_i(k) A_i x(k) + \sum_{i=1}^{q_1} \alpha_i(k) \sum_{j=1}^{q_2} \beta_j(k) B_j u(k)$$

$$= \sum_{i=1}^{q_1} \alpha_i(k) \left\{ A_i x(k) + \sum_{j=1}^{q_2} \beta_j(k) B_j u(k) \right\}$$

$$= \sum_{i=1}^{q_1} \alpha_i(k) \left\{ \sum_{j=1}^{q_2} \beta_j(k) A_i x(k) + \sum_{j=1}^{q_2} \beta_j(k) B_j u(k) \right\}$$

$$= \sum_{i=1}^{q_1} \alpha_i(k) \sum_{j=1}^{q_2} \beta_j(k) \left\{ A_i x(k) + B_j u(k) \right\}$$

$$= \sum_{i=1}^{q_1} \sum_{j=1}^{q_2} \alpha_i(k) \beta_j(k) \left\{ A_i x(k) + B_j u(k) \right\}$$

Consider the polytope Q_c whose vertices are obtained by taking all possible combinations of $\{A_i, B_j\}$ with $i = 1, 2, \ldots, q_1$ and $j = 1, 2, \ldots, q_2$. Since

$$\sum_{i=1}^{q_1} \sum_{j=1}^{q_2} \alpha_i(k) \beta_j(k) = \sum_{i=1}^{q_1} \alpha_i(k) \sum_{j=1}^{q_2} \beta_j(k) = 1$$

it follows that $\{A(k), B(k)\}$ can be expressed as a convex combination of the vertices of Q_c. $\qquad\square$

The state, the control and the disturbance are subject to the following polytopic constraints,

$$\begin{cases} x(k) \in X, & X = \left\{ x \in \mathbb{R}^n : F_x x \le g_x \right\} \\ u(k) \in U, & U = \left\{ u \in \mathbb{R}^m : F_u u \le g_u \right\} \\ w(k) \in W, & W = \left\{ w \in \mathbb{R}^d : F_w w \le g_w \right\} \end{cases} \tag{2.18}$$

where the matrices F_x, F_u, F_w and the vectors g_x, g_u, g_w are assumed to be constant with $g_x > 0$, $g_u > 0$, $g_w > 0$ such that the origin is contained in the interior of X, U and W. Here the inequalities are element-wise.

The aim of this section is to briefly review the set invariance theory, whose definitions are reported in the next subsection.

2.3.2 Basic Definitions

The relationship between the dynamic (2.15) and constraints (2.18) leads to the introduction of invariance/viability concepts [9, 28]. First we consider the case when no inputs are present,

$$x(k+1) = A(k)x(k) + Dw(k) \tag{2.19}$$

Definition 2.26 (Robustly positively invariant set) [24, 67] The set $\Omega \subseteq X$ is robustly positively invariant for system (2.19) if and only if, $\forall x(k) \in \Omega$, $\forall w(k) \in W$,

$$x(k+1) = A(k)x(k) + Dw(k) \in \Omega$$

Hence if the state $x(k)$ reaches Ω, it will remain inside Ω in spite of disturbance $w(k)$. The term *positively* refers to the fact that only forward evolutions of system (2.19) are considered and will be omitted in future sections for brevity.

The maximal robustly invariant set $\Omega_{max} \subseteq X$ is a robustly invariant set, that contains all the robustly invariant sets contained in X.

A concept similar to invariance, but with possibly stronger requirements, is the concept of contractivity introduced in the following definition.

Definition 2.27 (Robustly contractive set) [24, 67] For a given $0 \le \lambda \le 1$, the set $\Omega \subseteq X$ is robustly λ-contractive for system (2.19) if and only if, $\forall x(k) \in \Omega$, $\forall w(k) \in W$,

$$x(k+1) = A(k)x(k) + Dw(k) \in \lambda\Omega$$

Other useful definitions which will be used in the sequence are reported next.

Definition 2.28 (Robustly controlled invariant set and admissible control) [24, 67] The set $C \subseteq X$ is robustly controlled invariant for the system (2.15) if for all $x(k) \in C$, there exists a control value $u(k) \in U$ such that, $\forall w(k) \in W$,

$$x(k+1) = A(k)x(k) + B(k)u(k) + Dw(k) \in C$$

Such a control action is called *admissible*.

The maximal robustly controlled invariant set $C_{max} \subseteq X$ is a robustly controlled invariant set and contains all the robust controlled invariant sets contained in X.

Definition 2.29 (Robustly controlled contractive set) [24, 67] For a given $0 \le \lambda \le 1$ the set $C \subseteq X$ is robustly controlled contractive for the system (2.15) if for all $x(k) \in C$, there exists a control value $u(k) \in U$ such that, $\forall w(k) \in W$,

$$x(k+1) = A(k)x(k) + B(k)u(k) + Dw(k) \in \lambda C$$

Note that in Definition 2.27 and Definition 2.29 if $\lambda = 1$ we will, respectively, retrieve the robustly invariance and robustly controlled invariance.

2.3.3 Ellipsoidal Invariant Sets

In this subsection, ellipsoids will be used for set invariance description. For simplicity, the case of vanishing disturbances is considered. In other words, the system under consideration is,

$$x(k+1) = A(k)x(k) + B(k)u(k) \tag{2.20}$$

It is assumed that the state and control constraints are symmetric,

$$\begin{cases} x(k) \in X, X = \{x : |F_i^T x| \le 1\}, & \forall i = 1, 2, \ldots, n_1 \\ u(k) \in U, U = \{u : |u_j| \le u_{j\max}\}, & \forall j = 1, 2, \ldots, m \end{cases} \tag{2.21}$$

where $u_{j\max}$ is the j-component of vector $u_{\max} \in \mathbb{R}^m$.

Consider now the problem of checking robustly controlled invariance. The set $E(P) = \{x \in \mathbb{R}^n : x^T P^{-1} x \le 1\}$ is controlled invariant if and only if for all $x \in E(P)$ there exists an input $u = \Phi(x) \in U$ such that,

$$\left(A_i x + B_i \Phi(x)\right)^T P^{-1} \left(A_i x + B_i \Phi(x)\right) \le 1, \quad \forall i = 1, 2, \ldots, q \tag{2.22}$$

One possible choice for $u = \Phi(x)$ is a linear controller $u = Kx$. By defining $A_{ci} = A_i + B_i K$ with $i = 1, 2, \ldots, q$, condition (2.22) is equivalent to,

$$x^T A_{ci}^T P^{-1} A_{ci} x \le 1, \quad \forall i = 1, 2, \ldots, q \tag{2.23}$$

It is well known [29] that for the linear system (2.20), it is sufficient to check condition (2.23) for all x on the boundary of $E(P)$, i.e. for all x such that $x^T P^{-1} x = 1$. Therefore (2.23) can be transformed into,

$$x^T A_{ci}^T P^{-1} A_{ci} x \le x^T P^{-1} x, \quad \forall i = 1, 2, \ldots, q$$

or equivalently,

$$A_{ci}^T P^{-1} A_{ci} \preceq P^{-1}, \quad \forall i = 1, 2, \ldots, q$$

By using the Schur complements, this condition can be rewritten as,

$$\begin{bmatrix} P^{-1} & A_{ci}^T \\ A_{ci} & P \end{bmatrix} \succeq 0, \quad \forall i = 1, 2, \ldots, q$$

The condition provided here is not linear in P. By using the Schur complements again, one gets,

$$P - A_{ci} P A_{ci}^T \succeq 0, \quad \forall i = 1, 2, \ldots, q$$

or

$$\begin{bmatrix} P & A_{ci} P \\ P A_{ci}^T & P \end{bmatrix} \succeq 0, \quad \forall i = 1, 2, \ldots, q$$

By substituting $A_{ci} = A_i + B_i K, i = 1, 2, \ldots, q$, one obtains,

$$\begin{bmatrix} P & A_i P + B_i K P \\ P A_i^T + P K^T B_i^T & P \end{bmatrix} \succeq 0, \quad \forall i = 1, 2, \ldots, q \tag{2.24}$$

Though this condition is nonlinear, since P and K are the unknowns. Still it can be re-parameterized into a linear condition by setting $Y = KP$. Condition (2.24) becomes,

$$\begin{bmatrix} P & A_i P + B_i Y \\ P A_i^T + Y^T B_i^T & P \end{bmatrix} \succeq 0, \quad \forall i = 1, 2, \ldots, q \qquad (2.25)$$

Condition (2.25) is necessary and sufficient for ellipsoid $E(P)$ with the linear controller $u = Kx$ to be robustly invariant. Concerning the constraints (2.21), using equation (2.7) it follows that,

- The state constraints are satisfied if $E(P) \subseteq X$, or,

$$\begin{bmatrix} 1 & F_i^T P \\ P F_i & P \end{bmatrix} \succeq 0, \quad \forall i = 1, 2, \ldots, n_1 \qquad (2.26)$$

- The input constraints are satisfied if $E(P) \subseteq X_u$ where,

$$X_u = \{x \in \mathbb{R}^n : |K_j x| \leq u_{j\max}\}, \quad j = 1, 2, \ldots, m$$

and K_j is the j-row of the matrix $K \in \mathbb{R}^{m \times n}$, hence,

$$\begin{bmatrix} u_{j\max}^2 & K_j P \\ P K_j^T & P \end{bmatrix} \succeq 0,$$

Since $Y_j = K_j P$ where Y_j is the j-row of the matrix $Y \in \mathbf{R}^{m \times n}$, it follows that,

$$\begin{bmatrix} u_{j\max}^2 & Y_j \\ Y_j^T & P \end{bmatrix} \succeq 0 \qquad (2.27)$$

Define a vector $T_j \in \mathbb{R}^m$ as,

$$T_j = [0 \quad 0 \quad \ldots \quad 0 \quad \underbrace{1}_{j\text{-th position}} \quad 0 \quad \ldots \quad 0 \quad 0]$$

Since $Y_j = T_j Y$, equation (2.27) can be transformed into,

$$\begin{bmatrix} u_{j\max}^2 & T_j Y \\ Y^T T_j^T & P \end{bmatrix} \succeq 0 \qquad (2.28)$$

It is generally desirable to have the *largest* ellipsoid among the ones satisfying conditions (2.25), (2.26), (2.28). In the literature [55, 125], the size of ellipsoid $E(P)$ is usually measured by the determinant or the trace of matrix P. Here the trace of matrix P is chosen due to its linearity. The trace of a square matrix is defined to be the sum of the elements on the main diagonal of the matrix. Maximization of the trace of matrices corresponds to the search for the maximal sum of eigenvalues of matrices. With the trace of matrix as the objective function, the problem of choosing the *largest* robustly invariant ellipsoid can be formulated as,

$$J = \max_{P,Y}\{\text{trace}(P)\} \tag{2.29}$$

subject to

- Invariance condition (2.25).
- Constraints satisfaction (2.26), (2.28).

It is clear that the solution P, Y of problem (2.29) may lead to the controller $K = YP^{-1}$ such that the closed loop system with matrix $A_c(k) = A(k) + B(k)K$ is at the stability margin. In other words, the ellipsoid $E(P)$ thus obtained might not be contractive (although being invariant). Indeed, the system trajectories might not converge to the origin. In order to ensure $x(k) \to 0$ as $k \to \infty$, it is required that for all $x \in E(P)$, to have

$$\left(A_i x + B_i \Phi(x)\right)^T P^{-1}\left(A_i x + B_i \Phi(x)\right) < 1, \quad \forall i = 1, 2, \ldots, q$$

With the same argument as above, one can conclude that an ellipsoid $E(P)$ with a linear controller $u = Kx$ is robustly contractive if the following set of LMI conditions is satisfied,

$$\begin{bmatrix} P & A_i P + B_i Y \\ PA_i^T + Y^T B_i^T & P \end{bmatrix} \succ 0, \quad \forall i = 1, 2, \ldots, q \tag{2.30}$$

where $Y = KP$.

It should be noted that condition (2.30) is the same as (2.25), except that condition (2.30) requires the left hand side to be a strictly positive matrix.

2.3.4 Polyhedral Invariant Sets

The problem of set invariance description using polyhedral sets is addressed in this subsection. With linear constraints on the state and the control vectors, polyhedral invariant sets are preferred to ellipsoidal invariant sets, since they offer a better approximation of the domain of attraction [22, 38, 54]. To begin, let us consider the case, when the control input is of the form $u(k) = Kx(k)$. Then the system (2.15) becomes,

$$x(k + 1) = A_c(k)x(k) + Dw(k) \tag{2.31}$$

where

$$A_c(k) = A(k) + B(k)K = \text{Conv}\{A_{ci}\} \tag{2.32}$$

with $A_{ci} = A_i + B_i K$, $i = 1, 2, \ldots, q$.

The state constraints of the system (2.31) are,

$$x \in X_c, \quad X_c = \left\{ x \in \mathbb{R}^n : F_c x \le g_c \right\} \tag{2.33}$$

where

$$F_c = \begin{bmatrix} F_x \\ F_u K \end{bmatrix}, \qquad g_c = \begin{bmatrix} g_x \\ g_u \end{bmatrix}$$

The following definition plays an important role in computing robustly invariant sets for system (2.31) with constraints (2.33).

Definition 2.30 (Pre-image set) For the system (2.31), the one step admissible *pre-image set* of the set X_c is a set $X_c^{(1)} \subseteq X_c$ such that for all $x \in X_c^{(1)}$, it holds that, $\forall w \in W$,

$$A_{ci} x + D w \in X_c, \quad \forall i = 1, 2, \ldots, q$$

The pre-image set $X_c^{(1)}$ can be computed as [23, 27],

$$X_c^{(1)} = \left\{ x \in X_c : F_c A_{ci} x \le g_c - \max_{w \in W} \{ F_c D w \}, \ i = 1, 2, \ldots, q \right\} \tag{2.34}$$

Example 2.1 Consider the following uncertain and time-varying system,

$$x(k+1) = A(k)x(k) + Bu(k) + Dw(k)$$

where

$$A(k) = \alpha(k) A_1 + \left(1 - \alpha(k) \right) A_2, \quad 0 \le \alpha(k) \le 1,$$

$$A_1 = \begin{bmatrix} 1.1 & 1 \\ 0 & 1 \end{bmatrix}, \qquad A_2 = \begin{bmatrix} 0.6 & 1 \\ 0 & 1 \end{bmatrix}, \qquad B = \begin{bmatrix} 0 \\ 1 \end{bmatrix}, \qquad D = \begin{bmatrix} 1 & 0 \\ 0 & 1 \end{bmatrix}$$

The constraints (2.18) have the particular realization given by the matrices,

$$F_x = \begin{bmatrix} 1 & 0 \\ 0 & 1 \\ -1 & 0 \\ 0 & -1 \end{bmatrix}, \qquad g_x = \begin{bmatrix} 3 \\ 3 \\ 3 \\ 3 \end{bmatrix}, \qquad F_w = \begin{bmatrix} 1 & 0 \\ 0 & 1 \\ -1 & 0 \\ 0 & -1 \end{bmatrix}, \qquad g_w = \begin{bmatrix} 0.2 \\ 0.2 \\ 0.2 \\ 0.2 \end{bmatrix},$$

$$F_u = \begin{bmatrix} 1 \\ -1 \end{bmatrix}, \qquad g_u = \begin{bmatrix} 2 \\ 2 \end{bmatrix}$$

The controller is chosen as,

$$K = [-0.3856 \quad -1.0024]$$

The closed loop matrices are,

$$A_{c1} = \begin{bmatrix} 1.1000 & 1.0000 \\ -0.3856 & -0.0024 \end{bmatrix}, \qquad A_{c2} = \begin{bmatrix} 0.6000 & 1.0000 \\ -0.3856 & -0.0024 \end{bmatrix}$$

The set $X_c = \{x \in \mathbb{R}^2 : F_c x \le g_c\}$ is,

$$F_c = \begin{bmatrix} 1.0000 & 0 \\ 0 & 1.0000 \\ -1.0000 & 0 \\ 0 & -1.0000 \\ -0.3856 & -1.0024 \\ 0.3856 & 1.0024 \end{bmatrix}, \qquad g_c = \begin{bmatrix} 3.0000 \\ 3.0000 \\ 3.0000 \\ 3.0000 \\ 2.0000 \\ 2.0000 \end{bmatrix}$$

By solving the LP problem (2.13), it follows that the second and the fourth inequalities of X_c, i.e. $[0\ 1]x \le 3$ and $[0\ -1]x \le 3$, are redundant. After eliminating the redundant inequalities and normalizing the half-space representation, the set X_c is given as,

$$X_c = \left\{ x \in \mathbb{R}^2 : \widehat{F}_c x \le \widehat{g}_c \right\}$$

where

$$\widehat{F}_c = \begin{bmatrix} 1.0000 & 0 \\ -1.0000 & 0 \\ -0.3590 & -0.9333 \\ 0.3590 & 0.9333 \end{bmatrix}, \qquad \widehat{g}_c = \begin{bmatrix} 3.0000 \\ 3.0000 \\ 0.9311 \\ 0.9311 \end{bmatrix}$$

Using (2.34), the one step admissible pre-image set X_c^1 of X_c is defined as,

$$X_c^{(1)} = \left\{ x \in \mathbb{R}^2 : \begin{bmatrix} \widehat{F}_c \\ \widehat{F}_c A_1 \\ \widehat{F}_c A_2 \end{bmatrix} x \le \begin{bmatrix} \widehat{g}_c \\ \widehat{g}_c - \max_{w \in W}\{\widehat{F}_c w\} \\ \widehat{g}_c - \max_{w \in W}\{\widehat{F}_c w\} \end{bmatrix} \right\} \tag{2.35}$$

After removing redundant inequalities, the set $X_c^{(1)}$ is presented in minimal normalized half-space representation as,

$$X_c^{(1)} = \left\{ x \in \mathbb{R}^2 : \begin{bmatrix} 1.0000 & 0 \\ -1.0000 & 0 \\ -0.3590 & -0.9333 \\ 0.3590 & 0.9333 \\ 0.7399 & 0.6727 \\ -0.7399 & -0.6727 \\ 0.3753 & -0.9269 \\ -0.3753 & 0.9269 \end{bmatrix} x \le \begin{bmatrix} 3.0000 \\ 3.0000 \\ 0.9311 \\ 0.9311 \\ 1.8835 \\ 1.8835 \\ 1.7474 \\ 1.7474 \end{bmatrix} \right\}$$

The sets X, X_c and $X_c^{(1)}$ are depicted in Fig. 2.4.

Procedure 2.1 Robustly invariant set computation [46, 68]
- **Input:** The matrices $A_{c1}, A_{c2}, \ldots, A_{cq}, D$, the sets X_c in (2.33) and W.
- **Output:** The robustly invariant set Ω.

1. Set $i = 0$, $F_0 = F_c$, $g_0 = g_c$ and $X_0 = \{x \in \mathbb{R}^n : F_0 x \le g_0\}$.
2. Set $X_1 = X_0$.
3. Eliminate redundant inequalities of the following polytope,

$$
P = \left\{ x \in \mathbb{R}^n : \begin{bmatrix} F_0 \\ F_0 A_{c1} \\ F_0 A_{c2} \\ \vdots \\ F_0 A_{cq} \end{bmatrix} x \le \begin{bmatrix} g_0 \\ g_0 - \max_{w \in W}\{F_0 D w\} \\ g_0 - \max_{w \in W}\{F_0 D w\} \\ \vdots \\ g_0 - \max_{w \in W}\{F_0 D w\} \end{bmatrix} \right\}
$$

4. Set $X_0 = P$ and update consequently the matrices F_0 and g_0.
5. If $X_0 = X_1$ then stop and set $\Omega = X_0$. Else continue.
6. Set $i = i + 1$ and go to step 2.

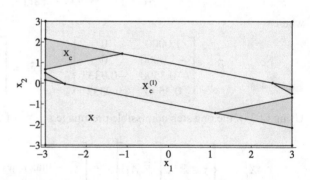

Fig. 2.4 One step pre-image set for Example 2.1

Clearly, $\Omega \subseteq X_c$ is robustly invariant if it equals to its one step admissible pre-image set, i.e., $\forall x \in \Omega$, $\forall w \in W$,

$$
A_i x + D w \in \Omega, \quad \forall i = 1, 2, \ldots, q
$$

Using this observation, Procedure 2.1 can be used for computing the set Ω for system (2.31) with constraints (2.33).

A natural question for Procedure 2.1 is that if there exists a finite index i such that $X_0 = X_1$, or equivalently if Procedure 2.1 terminates after a finite number of iterations.

In the absence of disturbances, the following theorem holds [25].

Theorem 2.2 [25] *Assume that the system* (2.31) *is robustly asymptotically stable. Then there exists a finite index* $i = i_{\max}$, *such that* $X_0 = X_1$ *in Procedure 2.1.*

Procedure 2.2 Robustly invariant set computation

- **Input:** The matrices $A_{c1}, A_{c2}, \ldots, A_{cq}, D$, the sets X_c and W.
- **Output:** The robustly invariant set Ω.

1. Set $i = 0$, $F_0 = F_c$, $g_0 = g_c$ and $X_0 = \{x \in \mathbb{R}^n : F_0 x \leq g_0\}$.
2. Consider the following polytope

$$
P = \left\{ x \in \mathbb{R}^n : \begin{bmatrix} F_0 \\ F_0 A_{c1} \\ F_0 A_{c2} \\ \vdots \\ F_0 A_{cq} \end{bmatrix} x \leq \begin{bmatrix} g_0 \\ g_0 - \max_{w \in W}\{F_0 D w\} \\ g_0 - \max_{w \in W}\{F_0 D w\} \\ \vdots \\ g_0 - \max_{w \in W}\{F_0 D w\} \end{bmatrix} \right\}
$$

and iteratively check the redundancy of the following subset of inequalities

$$
\left\{ x \in \mathbb{R}^n : F_0 A_{cj} x \leq g_0 - \max_{w \in W}\{F_0 D w\} \right\}
$$

   with $j = 1, 2, \ldots, q$.
3. If all of the inequalities are redundant with respect to X_0, then stop and set $\Omega = X_0$. Else continue.
4. Set $X_0 = P$.
5. Set $i = i + 1$ and go to step 2.

Remark 2.1 In the presence of disturbances, the necessary and sufficient condition for the existence of a finite index i is that the minimal robustly invariant set[3] [71, 97, 105] is a subset of X_c.

Note that checking equality of two polytopes X_0 and X_1 in step 5 is computationally demanding, i.e. one has to check $X_0 \subseteq X_1$ and $X_1 \subseteq X_0$. Note also that if the set Ω is invariant at the iteration i then the following set of inequalities,

$$
\begin{bmatrix} F_0 A_{c1} \\ F_0 A_{c2} \\ \vdots \\ F_0 A_{cq} \end{bmatrix} x \leq \begin{bmatrix} g_0 - \max_{w \in W}\{F_0 D_1 w\} \\ g_0 - \max_{w \in W}\{F_0 D_2 w\} \\ \vdots \\ g_0 - \max_{w \in W}\{F_0 D_q w\} \end{bmatrix}
$$

is redundant with respect to $\Omega = \{x \in \mathbb{R}^n : F_0 x \leq g_0\}$. Hence Procedure 2.1 can be made more efficient as in Procedure 2.2.

---

[3]The set $\Omega \subseteq X_c$ is minimal robustly invariant if it is a robustly invariant set and is a subset of any robustly invariant set contained in X_c.

Fig. 2.5 Maximal robustly
invariant set Ω_{max} for
Example 2.2

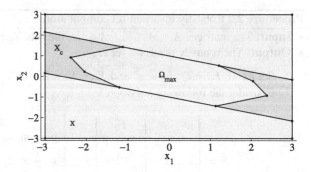

It is well known [29, 46, 71] that the set Ω resulting from Procedure 2.1 or
Procedure 2.2, is actually the maximal robustly invariant set for system (2.31) and
constraints (2.33), that is $\Omega = \Omega_{max}$.

Example 2.2 Consider the uncertain system in Example 2.1 with the same state,
control and disturbance constraints. Using Procedure 2.2, the set Ω_{max} is found
after 5 iterations as,

$$\Omega_{max} = \left\{ x \in \mathbb{R}^2 : \begin{bmatrix} -0.3590 & -0.9333 \\ 0.3590 & 0.9333 \\ 0.6739 & 0.7388 \\ -0.6739 & -0.7388 \\ 0.8979 & 0.4401 \\ -0.8979 & -0.4401 \\ 0.3753 & -0.9269 \\ -0.3753 & 0.9269 \end{bmatrix} x \leq \begin{bmatrix} 0.9311 \\ 0.9311 \\ 1.2075 \\ 1.2075 \\ 1.7334 \\ 1.7334 \\ 1.7474 \\ 1.7474 \end{bmatrix} \right\}$$

The sets X, X_c and Ω_{max} are depicted in Fig. 2.5.

Definition 2.31 (One-step robustly controlled set) Given system (2.15), the one-
step robustly controlled set, denoted as C_1 of the set $C_0 = \{x \in \mathbb{R}^n : F_0 x \leq g_0\}$ is
given by all states that can be steered in one step into C_0 when a suitable control
action is applied. The set C_1 can be shown to be [23, 27],

$$C_1 = \left\{ x \in \mathbb{R}^n : \exists u \in U : F_0(A_i x + B_i u) \leq g_0 - \max_{w \in W}\{F_0 D w\} \right\}, \quad i = 1, 2, \ldots, q$$
(2.36)

Remark 2.2 If C_0 is robustly invariant, then $C_0 \subseteq C_1$. Hence C_1 is a robustly controlled invariant set.

Recall that Ω_{max} is the maximal robustly invariant set with respect to a predefined
control law $u(k) = Kx(k)$. Define C_N as the set of all states, that can be steered into
Ω_{max} in no more than N steps along an admissible trajectory, i.e. a trajectory satisfy-

Fig. 2.6 Robustly controlled invariant sets for Example 2.3

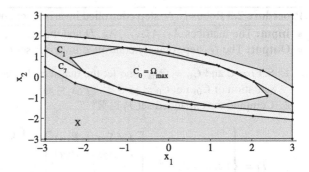

ing control, state and disturbance constraints. This set can be computed recursively by Procedure 2.3.

Since Ω_{max} is a robustly invariant set, it follows that $C_{i-1} \subseteq C_i$ and therefore C_i is a robustly controlled invariant set and a sequence of nested polytopes. Note that the complexity of C_N does not have an analytic dependence on N and may increase without bound, thus placing a practical limitation on the choice of N.

Example 2.3 Consider the uncertain system in Example 2.1 with the same state, input and disturbance constraints. Using Procedure 2.3, the robustly controlled invariant sets C_N with $N = 1$ and $N = 7$ are obtained and shown in Fig. 2.6. Note that $C_7 = C_8$ is the maximal robustly controlled invariant set.

The set C_7 is presented in minimal normalized half-space representation as,

$$
C_7 = \left\{ x \in \mathbb{R}^2 :
\begin{bmatrix}
0.3731 & 0.9278 \\
-0.3731 & -0.9278 \\
0.4992 & 0.8665 \\
-0.4992 & -0.8665 \\
0.1696 & 0.9855 \\
-0.1696 & -0.9855 \\
0.2142 & 0.9768 \\
-0.2142 & -0.9768 \\
0.7399 & 0.6727 \\
-0.7399 & -0.6727 \\
1.0000 & 0 \\
-1.0000 & 0
\end{bmatrix}
x \leq
\begin{bmatrix}
1.3505 \\
1.3505 \\
1.3946 \\
1.3946 \\
1.5289 \\
1.5289 \\
1.4218 \\
1.4218 \\
1.8835 \\
1.8835 \\
3.0000 \\
3.0000
\end{bmatrix}
\right\}
$$

2.4 On the Domain of Attraction

In this section we study the problem of estimating the domain of attraction for uncertain and/or time-varying linear discrete-time systems in closed-loop with a *saturated* controller and state constraints.

Procedure 2.3 Robustly N-step controlled invariant set computation

- **Input:** The matrices $A_1, A_2, \ldots, A_q, D$ and the sets X, U, W and $\Omega_{\max}$.
- **Output:** The N-step robustly controlled invariant set C_N.

1. Set $i = 0$ and $C_0 = \Omega_{\max}$ and let the matrices F_0, g_0 be the half-space representation of C_0, i.e. $C_0 = \{x \in \mathbb{R}^n : F_0 x \leq g_0\}$
2. Compute the expanded set $P_i \subset \mathbb{R}^{n+m}$

$$P_i = \left\{ (x, u) \in \mathbb{R}^{n+m} : \begin{bmatrix} F_i(A_1 x + B_1 u) \\ F_i(A_2 x + B_2 u) \\ \vdots \\ F_i(A_q x + B_q u) \end{bmatrix} \leq \begin{bmatrix} g_i - \max_{w \in W}\{F_i D w\} \\ g_i - \max_{w \in W}\{F_i D w\} \\ \vdots \\ g_i - \max_{w \in W}\{F_i D w\} \end{bmatrix} \right\}$$

3. Compute the projection $P_i^{(n)}$ of P_i on $\mathbb{R}^n$

$$P_i^{(n)} = \left\{ x \in \mathbb{R}^n : \exists u \in U \text{ such that } (x, u) \in P_i \right\}$$

4. Set

$$C_{i+1} = P_i^{(n)} \cap X$$

and let F_{i+1}, g_{i+1} be the half-space representation of C_{i+1}, i.e.

$$C_{i+1} = \left\{ x \in \mathbb{R}^n : F_{i+1} x \leq g_{i+1} \right\}$$

5. If $C_{i+1} = C_i$, then stop and set $C_N = C_i$. Else continue.
6. If $i = N$, then stop else continue.
7. Set $i = i + 1$ and go to step 2.

2.4.1 Problem Formulation

Consider the following uncertain and/or time-varying linear discrete-time system,

$$x(k + 1) = A(k)x(k) + B(k)u(k) \tag{2.37}$$

where

$$\begin{cases} A(k) = \sum_{i=1}^{q} \alpha_i(k) A_i, \qquad B(k) = \sum_{i=1}^{q} \alpha_i(k) B_i \\ \sum_{i=1}^{q} \alpha_i(k) = 1, \quad \alpha_i(k) \geq 0 \end{cases} \tag{2.38}$$

with given matrices $A_i \in \mathbb{R}^{n \times n}$ and $B_i \in \mathbb{R}^{n \times m}$, $i = 1, 2, \ldots, q$.

Fig. 2.7 Saturation function

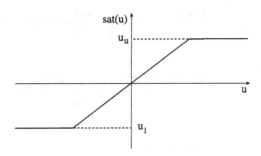

Both the state vector $x(k)$ and the control vector $u(k)$ are subject to the constraints,

$$\begin{cases} x(k) \in X, & X = \{x \in \mathbb{R}^n : F_i^T x \le g_i\}, & \forall i = 1, 2, \dots, n_1 \\ u(k) \in U, & U = \{u \in \mathbb{R}^m : u_{jl} \le u_j \le u_{ju}\}, & \forall j = 1, 2, \dots, m \end{cases} \tag{2.39}$$

where $F_i^T \in \mathbb{R}^n$ is the i-th row of the matrix $F_x \in \mathbb{R}^{n_1 \times n}$, g_i is the i-th component of the vector $g_x \in \mathbb{R}^{n_1}$, u_{il} and u_{iu} are respectively, the i-th component of the vectors u_l and u_u, which are the lower and upper bounds of input u. It is assumed that the matrix F_x and the vectors g_x, u_l and u_u are constant with $g_x > 0$, $u_l < 0$ and $u_u > 0$.

Our aim is to estimate the domain of attraction for the system,

$$x(k+1) = A(k)x(k) + B(k) \operatorname{sat}\big(Kx(k)\big) \tag{2.40}$$

subject to the constraints (2.39), where

$$u(k) = Kx(k) \tag{2.41}$$

is a given controller that robustly stabilizes system (2.37).

2.4.2 Saturation Nonlinearity Modeling—A Linear Differential Inclusion Approach

A linear differential inclusion approach used for modeling the saturation function is briefly reviewed in this subsection. This modeling framework was first proposed by Hu et al. [55, 57, 58]. Then its generalization was developed by Alamo et al. [5, 6]. The main idea of the linear differential inclusion approach is to use an auxiliary vector variable $v \in \mathbb{R}^m$, and to compose the output of the saturation function as a convex combination of u and v.

The saturation function is defined as

$$\operatorname{sat}(u) = \begin{bmatrix} \operatorname{sat}(u_1) & \operatorname{sat}(u_2) & \cdots & \operatorname{sat}(u_m) \end{bmatrix}^T \tag{2.42}$$

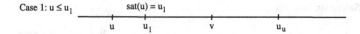

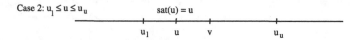

Fig. 2.8 Linear differential inclusion approach

where, see Fig. 2.7,

$$\text{sat}(u_i) = \begin{cases} u_{il}, & \text{if } u_i \leq u_{il} \\ u_i, & \text{if } u_{il} \leq u_i \leq u_{iu} \quad i = 1, 2, \ldots, m \\ u_{iu}, & \text{if } u_{iu} \leq u_i \end{cases} \tag{2.43}$$

To underline the details of the approach, let us first consider the case u and consequently v are scalars. Clearly, for any u, there exist $u_l \leq v \leq u_u$ and $0 \leq \beta \leq 1$ such that,

$$\text{sat}(u) = \beta u + (1 - \beta)v \tag{2.44}$$

or, equivalently

$$\text{sat}(u) \in \text{Conv}\{u, v\} \tag{2.45}$$

Figure 2.8 illustrates this fact.

Analogously, for $m = 2$ and v such that

$$\begin{cases} u_{1l} \leq v_1 \leq u_{1u} \\ u_{2l} \leq v_2 \leq u_{2u} \end{cases} \tag{2.46}$$

the saturation function can be expressed as,

$$\text{sat}(u) = \beta_1 \begin{bmatrix} u_1 \\ u_2 \end{bmatrix} + \beta_2 \begin{bmatrix} u_1 \\ v_2 \end{bmatrix} + \beta_3 \begin{bmatrix} v_1 \\ u_2 \end{bmatrix} + \beta_4 \begin{bmatrix} v_1 \\ v_2 \end{bmatrix} \tag{2.47}$$

with $\sum_{j=1}^{4} \beta_j = 1$, $\beta_j \geq 0$. Or, equivalently

$$\text{sat}(u) \in \text{Conv}\left\{ \begin{bmatrix} u_1 \\ u_2 \end{bmatrix}, \begin{bmatrix} u_1 \\ v_2 \end{bmatrix}, \begin{bmatrix} v_1 \\ u_2 \end{bmatrix}, \begin{bmatrix} v_1 \\ v_2 \end{bmatrix} \right\} \tag{2.48}$$

Define D_m as the set of $m \times m$ diagonal matrices whose diagonal elements are either 0 or 1. For example, if $m = 2$ then

$$D_2 = \left\{ \begin{bmatrix} 0 & 0 \\ 0 & 0 \end{bmatrix}, \begin{bmatrix} 1 & 0 \\ 0 & 0 \end{bmatrix}, \begin{bmatrix} 0 & 0 \\ 0 & 1 \end{bmatrix}, \begin{bmatrix} 1 & 0 \\ 0 & 1 \end{bmatrix} \right\}$$

There are 2^m elements in D_m. Denote each element of D_m as E_j, $j = 1, 2, \ldots, 2^m$ and define $E_j^- = I - E_j$. For example, if

$$E_1 = \begin{bmatrix} 0 & 0 \\ 0 & 0 \end{bmatrix}$$

then

$$E_1^- = \begin{bmatrix} 1 & 0 \\ 0 & 1 \end{bmatrix} - \begin{bmatrix} 0 & 0 \\ 0 & 0 \end{bmatrix} = \begin{bmatrix} 1 & 0 \\ 0 & 1 \end{bmatrix}$$

Clearly, if $E_j \in D_m$, then E_j^- is also in D_m. The generalization of the results (2.45) (2.48) is reported by the following lemma [55, 57, 58],

Lemma 2.1 [57] *Consider two vectors $u \in \mathbb{R}^m$ and $v \in \mathbb{R}^m$ such that $u_{il} \leq v_i \leq u_{iu}$ for all $i = 1, 2, \ldots, m$, then it holds that*

$$\text{sat}(u) \in \text{Conv}\{E_j u + E_j^- v\}, \quad j = 1, 2, \ldots, 2^m \tag{2.49}$$

Consequently, there exist $\beta_j \geq 0$ and $\sum_{j=1}^{2^m} \beta_j = 1$ such that,

$$\text{sat}(u) = \sum_{j=1}^{2^m} \beta_j \left(E_j u + E_j^- v \right)$$

2.4.3 The Ellipsoidal Set Approach

The aim of this subsection is twofold. First, we provide an invariance condition of ellipsoidal sets for uncertain and/or time-varying linear discrete-time systems with a saturated input and state constraints. This invariance condition is an extended version of the previously published results in [57] for the robust case. Secondly, we propose a method for computing a *saturated* controller $u(k) = \text{sat}(Kx(k))$ that makes a given invariant ellipsoid *contractive* with the maximal contraction factor. For simplicity, the case of bounds equal to $u_{\max}$ is considered, namely

$$-u_l = u_u = u_{\max}$$

and let us assume that the set X in (2.39) is symmetric and $g_i = 1, \forall i = 1, 2, \ldots, n_1$. Clearly, the latter assumption is nonrestrictive as long as, $\forall g_i > 0$

$$F_i^T x \le g_i \quad \Leftrightarrow \quad \frac{F_i^T}{g_i} x \le 1$$

For a given matrix $H \in \mathbb{R}^{m \times n}$, define X_c as the intersection between X and the polyhedral set $F(H, u_{max}) = \{x \in \mathbb{R}^n : |Hx| \le u_{max}\}$, i.e.

$$X_c = \left\{ x \in \mathbb{R}^n : \begin{bmatrix} F_x \\ H \\ -H \end{bmatrix} x \le \begin{bmatrix} 1 \\ u_{max} \\ u_{max} \end{bmatrix} \right\} \qquad (2.50)$$

We are now ready to state the main result of this subsection,

Theorem 2.3 *If there exist a positive definite matrix $P \in \mathbb{R}^{n \times n}$ and a matrix $H \in \mathbb{R}^{m \times n}$ such that, $\forall i = 1, 2, \ldots, q, \forall j = 1, 2, \ldots, 2^m$,*

$$\begin{bmatrix} P & \{A_i + B_i(E_j K + E_j^- H)\} P \\ P\{A_i + B_i(E_j K + E_j^- H)\}^T & P \end{bmatrix} \succeq 0, \qquad (2.51)$$

and $E(P) \subset X_c$, then the ellipsoid $E(P)$ is a robustly invariant set for system (2.40) with constraints (2.39).

Proof Assume that there exist P and H such that condition (2.51) is satisfied. Using Lemma 2.1 and by choosing $v = Hx$, it follows that,

$$\mathrm{sat}\big(Kx(k)\big) = \sum_{j=1}^{2^m} \beta_j(k)\big(E_j Kx(k) + E_j^- Hx(k)\big)$$

for all $x(k)$ such that $|Hx(k)| \le u_{max}$. Subsequently,

$$
\begin{aligned}
x(k+1) &= \sum_{i=1}^{q} \alpha_i(k) \left\{ A_i + B_i \sum_{j=1}^{2^m} \beta_j(k)\big(E_j K + E_j^- H\big) \right\} x(k) \\
&= \sum_{i=1}^{q} \alpha_i(k) \left\{ \sum_{j=1}^{2^m} \beta_j(k) A_i + B_i \sum_{j=1}^{2^m} \beta_j(k)\big(E_j K + E_j^- H\big) \right\} x(k) \\
&= \sum_{i=1}^{q} \alpha_i(k) \sum_{j=1}^{2^m} \beta_j(k)\{A_i + B_i\big(E_j K + E_j^- H\big)\} x(k) \\
&= \sum_{i=1}^{q} \sum_{j=1}^{2^m} \alpha_i(k)\beta_j(k)\{A_i + B_i\big(E_j K + E_j^- H\big)\} x(k) = A_c(k)x(k)
\end{aligned}
$$

where

$$A_c(k) = \sum_{i=1}^{q} \sum_{j=1}^{2^m} \alpha_i(k)\beta_j(k)\{A_i + B_i(E_j K + E_j^- H)\}$$

Since

$$\sum_{i=1}^{q} \sum_{j=1}^{2^m} \alpha_i(k)\beta_j(k) = \sum_{i=1}^{q} \alpha_i(k)\left\{\sum_{j=1}^{2^m} \beta_j(k)\right\} = 1$$

if follows that $A_c(k)$ belongs to the polytope P_c, whose vertices are obtained by taking all possible combinations of $A_i + B_i(E_j K + E_j^- H)$ with $i = 1, 2, \ldots, q$ and $j = 1, 2, \ldots, 2^m$.

The set $E(P) = \{x \in \mathbb{R}^n : x^T P^{-1} x \leq 1\}$ is invariant, if and only if

$$x^T A_c(k)^T P^{-1} A_c(k) x \leq 1 \tag{2.52}$$

for all $x \in E(P)$. With the same argument as in Sect. 2.3.3, condition (2.52) can be transformed to,

$$\begin{bmatrix} P & A_c(k)P \\ PA_c(k)^T & P \end{bmatrix} \succeq 0 \tag{2.53}$$

Since $A_c(k)$ belongs to the polytope P_c, it follows that one should check (2.53) at the vertices of P_c. So the set of LMI conditions to be satisfied is the following, $\forall i = 1, 2, \ldots, q, \forall j = 1, 2, \ldots, 2^m$,

$$\begin{bmatrix} P & \{A_i + B_i(E_j K + E_j^- H)\}P \\ P\{A_i + B_i(E_j K + E_j^- H)\}^T & P \end{bmatrix} \succeq 0 \qquad \square$$

Note that condition (2.51) involves the multiplication between two unknown parameters H and P. By defining $Y = HP$, condition (2.51) can be rewritten as, $\forall i = 1, 2, \ldots, q, \forall j = 1, 2, \ldots, 2^m$,

$$\begin{bmatrix} P & (A_i P + B_i E_j K P + B_i E_j^- Y) \\ (PA_i^T + PK^T E_j B_i^T + Y^T E_j^- B_i^T) & P \end{bmatrix} \succeq 0, \tag{2.54}$$

Thus the unknown matrices P and Y enter linearly in (2.54).

As in Sect. 2.3.3, in general one would like to have the largest invariant ellipsoid for system (2.37) under *saturated* controller $u(k) = \text{sat}(Kx(k))$ with respect to constraints (2.39). This can be achieved by solving the following LMI problem,

$$J = \max_{P,Y} \{\text{trace}(P)\} \tag{2.55}$$

subject to

- Invariance condition (2.54).
- Constraint satisfaction,

    – On state

$$\begin{bmatrix} 1 & F_i^T P \\ P F_i & P \end{bmatrix} \geq 0, \quad \forall i = 1, 2, \ldots, n_1$$

    – On input

$$\begin{bmatrix} u_{i\max}^2 & Y_i \\ Y_i^T & P \end{bmatrix} \geq 0, \quad \forall i = 1, 2, \ldots, m$$

    where Y_i is the i-th row of the matrix Y.

Example 2.4 Consider the following linear uncertain, time-varying discrete-time system,

$$x(k+1) = A(k)x(k) + B(k)u(k)$$

with

$$A(k) = \alpha(k)A_1 + \left(1 - \alpha(k)\right)A_2, \qquad B(k) = \alpha(k)B_1 + \left(1 - \alpha(k)\right)B_2$$

and

$$A_1 = \begin{bmatrix} 1 & 0.1 \\ 0 & 1 \end{bmatrix}, \qquad A_2 = \begin{bmatrix} 1 & 0.2 \\ 0 & 1 \end{bmatrix}, \qquad B_1 = \begin{bmatrix} 0 \\ 1 \end{bmatrix}, \qquad B_2 = \begin{bmatrix} 0 \\ 1.5 \end{bmatrix}$$

At each sampling time $\alpha(k) \in [0, 1]$ is an uniformly distributed pseudo-random number. The constraints are,

$$-10 \leq x_1 \leq 10, \qquad -10 \leq x_2 \leq 10, \qquad -1 \leq u \leq 1$$

The controller is chosen as,

$$K = [-1.8112 \quad -0.8092]$$

By solving the LMI problem (2.55), the matrices P and Y are obtained,

$$P = \begin{bmatrix} 5.0494 & -8.9640 \\ -8.9640 & 28.4285 \end{bmatrix}, \qquad Y = [0.4365 \quad -4.2452]$$

Hence

$$H = Y P^{-1} = [-0.4058 \quad -0.2773]$$

Solving the LMI problem (2.30), the ellipsoid $E(P_1)$ is obtained with

$$P_1 = \begin{bmatrix} 1.1490 & -3.1747 \\ -3.1747 & 9.9824 \end{bmatrix}$$

Fig. 2.9 Invariant sets with different control laws for Example 2.4. The set $E(P)$ is obtained for $u(k) = \text{sat}(Kx(k))$ and the set $E(P_1)$ is obtained for $u(k) = Kx(k)$

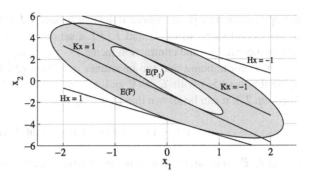

Fig. 2.10 State trajectories of the closed loop system for Example 2.4

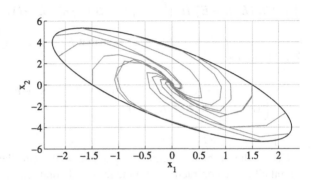

under the linear feedback $u(k) = Kx(k)$. Figure 2.9 presents two invariant ellipsoids with different control laws. $E(P)$ is obtained for $u(k) = \text{sat}(Kx(k))$ and $E(P_1)$ is obtained for $u(k) = Kx(k)$.

Figure 2.10 shows state trajectories of the closed loop system with the controller $u(k) = \text{sat}(Kx(k))$ for different initial conditions and realizations of $\alpha(k)$.

In the first part of this subsection, Theorem 2.3 was exploited in the following manner: if $E(P)$ is robustly invariant for the system,

$$x(k+1) = A(k)x(k) + B(k)\,\text{sat}\big(Kx(k)\big)$$

then there exists a stabilizing linear controller $u(k) = Hx(k)$, such that $E(P)$ is robustly invariant for system,

$$x(k+1) = A(k)x(k) + B(k)Hx(k)$$

with $H \in \mathbb{R}^{m \times n}$ obtained by solving the LMI problem (2.55).

Theorem 2.3 now will be exploited in a different manner. We would like to design a *saturated* controller $u(k) = \text{sat}(Kx(k))$ that makes a *given* invariant ellipsoid $E(P)$ contractive with the maximal contraction factor. This invariant ellipsoid $E(P)$ can be inherited for example together with a linear controller $u(k) = Hx(k)$ from the optimization of some convex objective function $J(P)$,[4] for example trace(P). In the second stage, using H and $E(P)$, a saturated controller $u(k) = \text{sat}(Kx(k))$ which maximizes some contraction factor $1 - g$ is computed.

It is worth noticing that the invariance condition (2.29) corresponds to the one in condition (2.54) with $E_j = 0$ and $E_j^- = I - E_j = I$. Following the proof of Theorem 2.3, it can be shown that for the system,

$$x(k+1) = A(k)x(k) + B(k)\,\text{sat}\big(Kx(k)\big)$$

the set $E(P)$ is contractive with the contraction factor $1 - g$ if

$$\big\{A_i + B_i\big(E_j K + E_j^- H\big)\big\}^T P^{-1}\big\{A_i + B_i\big(E_j K + E_j^- H\big)\big\} - P^{-1} \preceq -g P^{-1} \quad (2.56)$$

$\forall i = 1, 2, \ldots, q, \forall j = 1, 2, \ldots, 2^m$ such that $E_j \neq 0$. Using the Schur complements, (2.56) becomes,

$$\begin{bmatrix} (1-g)P^{-1} & (A_i + B_i(E_j K + E_j^- H))^T \\ (A_i + B_i(E_j K + E_j^- H)) & P \end{bmatrix} \succeq 0 \qquad (2.57)$$

$\forall i = 1, 2, \ldots, p, \forall j = 1, 2, \ldots, 2^m$ with $E_j \neq 0$.

Hence, the problem of computing a saturated controller that makes a given invariant ellipsoid contractive with the maximal contraction factor can be formulated as,

$$J = \max_{g, K}\{g\} \qquad (2.58)$$

subject to (2.57).

Recall that here the only unknown parameters are the matrix $K \in \mathbb{R}^{m \times n}$ and the scalar g, the matrices P and H being given in the first stage.

Remark 2.3 The proposed two-stage control design presented here benefits from global uniqueness properties of the solution. This is due to the one-way dependence of the two (prioritized) objectives: the trace maximization precedes the associated contraction factor.

Example 2.5 Consider the uncertain system in Example 2.4 with the same state and input constraints. In the first stage, by solving (2.29), the matrices P and Y are obtained,

$$P = \begin{bmatrix} 100.0000 & -43.1051 \\ -43.1051 & 100.0000 \end{bmatrix}, \qquad Y = [-3.5691 \quad -6.5121]$$

---

[4]Practically, the design of the invariant ellipsoid $E(P)$ and the controller $u(k) = Hx(k)$ can be done by solving the LMI problem (2.29).

Fig. 2.11 Invariant ellipsoid
and state trajectories of the
closed loop system for
Example 2.5

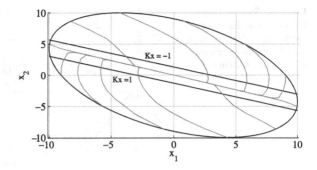

Hence $H = YP^{-1} = [-0.0783 \; -0.0989]$. In the second stage, by solving (2.58),
the matrix K is obtained,

$$K = [-0.3342 \quad -0.7629]$$

Figure 2.11 shows the invariant ellipsoid $E(P)$. This figure also shows state trajectories of the closed loop system with the controller $u(k) = \text{sat}(Kx(k))$ for different
initial conditions and realizations of $\alpha(k)$.

For the initial condition $x(0) = [-4 \; 10]^T$, Fig. 2.12(a) presents state trajectories
of the closed loop system as functions of time for the saturated controller $u(k) =
\text{sat}(Kx(k))$ (solid) and for the linear controller $u(k) = Hx(k)$ (dashed). It is worth
noticing that the time to regulate the plant to the origin by using $u(k) = Hx(k)$ is
longer than the time to regulate the plant to the origin by using $u(k) = \text{sat}(Kx(k))$.
The reason is that when using $u(k) = Hx(k)$, the control action is saturated only at
some points of the boundary of $E(P)$, while using $u(k) = \text{sat}(Kx(k))$, the control
action is saturated not only on the boundary of $E(P)$, the saturation being active
also inside $E(P)$. This phenomena can be observed in Fig. 2.12(b). The same figure
presents the realization of $\alpha(k)$.

2.4.4 The Polyhedral Set Approach

The problem of estimating the domain of attraction is addressed by using polyhedral
sets in this subsection. For a given linear controller $u(k) = Kx(k)$, it is clear that the
largest polyhedral invariant set is the maximal robustly invariant set $\Omega_{\max}$ which can
be found using Procedure 2.1 or Procedure 2.2. From this point on, it is assumed that
$\Omega_{\max}$ is known.

The aim is to find the *largest* polyhedral invariant set $\Omega_s \subseteq X$ characterizing an
estimation of the domain of attraction for system (2.37) under $u(k) = \text{sat}(Kx(k))$.
To this aim, recall that from Lemma 2.1, the saturation function can be expressed
as,

$$\text{sat}(Kx(k)) = \sum_{j=1}^{2^m} \beta_j(k)(E_j Kx + E_j^- v), \qquad \sum_{j=1}^{2^m} \beta_j(k) = 1, \quad \beta_j \geq 0 \quad (2.59)$$

Fig. 2.12 State and input trajectories for Example 2.5 for the controller $u = \text{sat}(Kx)$ (*solid*), and for the controller $u = Hx$ (*dashed*) in the figures for x_1, x_2 and u

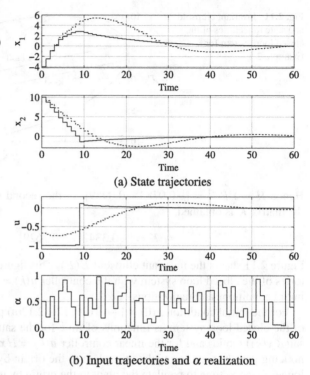

(a) State trajectories

(b) Input trajectories and α realization

where $u_l \leq v \leq u_u$ and E_j is an element of D_m and $E_j^- = I - E_j$.

Using (2.59), the closed loop system can be rewritten as,

$$x(k+1) = \sum_{i=1}^{q} \alpha_i(k) \left\{ A_i x(k) + B_i \sum_{j=1}^{2^m} \beta_j(k) \left(E_j K x(k) + E_j^- v \right) \right\}$$

$$= \sum_{i=1}^{q} \alpha_i(k) \left\{ \sum_{j=1}^{2^m} \beta_j(k) A_i x(k) + B_i \sum_{j=1}^{2^m} \beta_j(k) \left(E_j K x(k) + E_j^- v \right) \right\}$$

$$= \sum_{i=1}^{q} \alpha_i(k) \sum_{j=1}^{2^m} \beta_j(k) \left\{ A_i x(k) + B_i \left(E_j K x(k) + E_j^- v \right) \right\}$$

or

$$x(k+1) = \sum_{i=1}^{q} \alpha_i(k) \sum_{j=1}^{2^m} \beta_j(k) \left\{ (A_i + B_i E_j K) x(k) + B_i E_j^- v \right\} \qquad (2.60)$$

The variable $v \in \mathbb{R}^m$ can be considered as an external controlled input for the system (2.60). Hence, the problem of finding Ω_s for the system (2.40) boils down to the problem of computing the largest controlled invariant set for the system (2.60).

Procedure 2.4 Invariant set computation

- **Input:** The matrices $A_1, \ldots, A_q, B_1, \ldots, B_q$, the matrix K and the sets X, U and $\Omega_{\max}$
- **Output:** An invariant approximation of the invariant set Ω_s for the closed loop system (2.40).

  1. Set $i = 0$ and $C_0 = \Omega_{\max}$ and let the matrices F_0, g_0 be the half space representation of C_0, i.e. $C_0 = \{x \in \mathbb{R}^n : F_0 x \leq g_0\}$
  2. Compute the expanded set $P_j \subset \mathbb{R}^{n+m}, \forall j = 1, 2, \ldots, 2^m$

$$
P_j = \left\{ (x, v) \in \mathbb{R}^{n+m} : \begin{bmatrix} F_i\{(A_1 + B_1 E_j K)x + B_1 E_j^- v\} \\ F_i\{(A_2 + B_2 E_j K)x + B_2 E_j^- v\} \\ \vdots \\ F_i\{(A_q + B_q E_j K)x + B_q E_j^- v\} \end{bmatrix} \leq \begin{bmatrix} g_i \\ g_i \\ \vdots \\ g_i \end{bmatrix} \right\}
$$

  3. Compute the projection $P_j^{(n)}$ of P_j on $\mathbb{R}^n$,

$$
P_j^{(n)} = \{x \in \mathbb{R}^n : \exists v \in U \text{ such that } (x, v) \in P_j\}, \quad \forall j = 1, 2, \ldots, 2^m
$$

  4. Set

$$
C_{i+1} = X \bigcap_{j=1}^{2^m} P_j^{(n)}
$$

  and let the matrices F_{i+1}, g_{i+1} be the half space representation of C_{i+1}, i.e.

$$
C_{i+1} = \{x \in \mathbb{R}^n : F_{i+1} x \leq g_{i+1}\}
$$

  5. If $C_{i+1} = C_i$, then stop and set $\Omega_s = C_i$. Else continue.
  6. Set $i = i + 1$ and go to step 2.

The system (2.60) can be considered as an uncertain system with respect to the parameters α_i and β_j. Hence using the results in Sect. 2.3.4, Procedure 2.4 can be used to obtain Ω_s.

Since $\Omega_{\max}$ is robustly invariant, it follows that $C_{i-1} \subseteq C_i$. Hence C_i is a robustly invariant set. The set sequence $\{C_0, C_1, \ldots\}$ converges to Ω_s, which is the largest polyhedral invariant set.

Remark 2.4 Each polytope C_i represents an inner invariant approximation of the domain of attraction for the system (2.37) under the controller $u(k) = \text{sat}(Kx(k))$. That means Procedure 2.4 can be stopped at any time before converging to the true largest invariant set Ω_s and obtain an inner invariant approximation of the domain of attraction.

Procedure 2.5 Invariant set computation

- **Input:** The matrices $A_1, A_2, \ldots, A_q$ and the sets X_H and $\Omega_{\max}$.
- **Output:** The invariant set Ω_s^H.

1. Set $i = 0$ and $C_0 = \Omega_{\max}$ and let the matrices F_0, g_0 be the half-space representation of the set C_0, i.e. $C_0 = \{x \in \mathbb{R}^n : F_0 x \le g_0\}$
2. Compute the set $P_j \subset \mathbb{R}^n$

$$P_j = \left\{ x \in \mathbb{R}^n : \begin{bmatrix} F_i(A_1 + B_1 E_j K + B_1 E_j^- H)x \\ F_i(A_2 + B_2 E_j K + B_2 E_j^- H)x \\ \vdots \\ F_i(A_q + B_q E_j K + + B_q E_j^- H)x \end{bmatrix} \le \begin{bmatrix} g_i \\ g_i \\ \vdots \\ g_i \end{bmatrix} \right\}$$

3. Set

$$C_{i+1} = X_H \bigcap_{j=1}^{2^m} P_j$$

and let the matrices F_{i+1}, g_{i+1} be the half-space representation of C_{i+1}, i.e.

$$C_{i+1} = \left\{ x \in \mathbb{R}^n : F_{i+1} x \le g_{i+1} \right\}$$

4. If $C_{i+1} = C_i$, then stop and set $\Omega_s = C_i$. Else continue.
5. Set $i = i + 1$ and go to step 2.

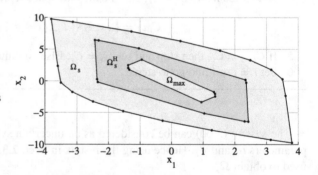

Fig. 2.13 Invariant sets for different control laws and different methods for Example 2.6. The set Ω_s is obtained for $u(k) = \text{sat}(Kx(k))$ using Procedure 2.4. The set Ω_s^H is obtained for $u(k) = \text{sat}(Kx(k))$ using Procedure 2.5. The set $\Omega_{\max}$ is obtained for $u(k) = Kx$ using Procedure 2.2

It is worth noticing that the matrix $H \in \mathbb{R}^{m \times n}$ resulting from the LMI problem (2.55) can also be used for computing an inner polyhedral invariant approximation Ω_s^H of the domain of attraction. Clearly, Ω_s^H is a subset of Ω_s, since v is now in a restricted form $v(k) = Hx(k)$. In this case, using (2.60) one obtains,

$$x(k+1) = \sum_{i=1}^{q} \alpha_i(k) \sum_{j=1}^{2^m} \beta_j(k) \left\{ \left(A_i + B_i E_j K + B_i E_j^- H \right) x(k) \right\} \qquad (2.61)$$

Fig. 2.14 State trajectories of the closed loop system with $u(k) = \text{sat}(Kx(k))$ for Example 2.6

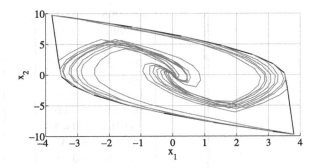

Define the set X_H as,

$$X_H = \{x \in \mathbb{R}^n : F_H x \leq g_H\} \tag{2.62}$$

where

$$F_H = \begin{bmatrix} F_x \\ H \\ -H \end{bmatrix}, \qquad g_H = \begin{bmatrix} g_x \\ u_u \\ u_l \end{bmatrix}$$

Procedure 2.5 can be used for computing Ω_s^H.

Since the matrix $\sum_{i=1}^{q} \alpha_i(k) \sum_{j=1}^{2m} \beta_j(k)\{(A_i + B_i E_j K + B_i E_j^- H)\}$ is asymptotically stable, Procedure 2.5 terminates in finite time [29]. In other words, there exists a finite index $i = i_{\max}$ such that $C_{i_{\max}} = C_{i_{\max}+1}$.

Example 2.6 Consider Example 2.4 with the same state and control constraints. The controller is $K = [-1.8112 \; -0.8092]$.

Using Procedure 2.4, the set Ω_s is obtained after 121 iterations and depicted in Fig. 2.13. This figure also shows the set Ω_s^H obtained by using Procedure 2.5 with the auxiliary matrix $H = [-0.4058 \; -0.2773]$, and the set $\Omega_{\max}$ obtained with the controller $u(k) = Kx$ using Procedure 2.2.

Ω_s^H and Ω_s are presented in minimal normalized half-space representation as,

$$\Omega_s^H = \left\{ x \in \mathbb{R}^2 : \begin{bmatrix} -0.8256 & -0.5642 \\ 0.8256 & 0.5642 \\ 0.9999 & 0.0108 \\ -0.9999 & -0.0108 \\ 0.9986 & 0.0532 \\ -0.9986 & -0.0532 \\ -0.6981 & -0.7160 \\ 0.6981 & 0.7160 \\ 0.9791 & 0.2033 \\ -0.9791 & -0.2033 \\ -0.4254 & -0.9050 \\ 0.4254 & 0.9050 \end{bmatrix} x \leq \begin{bmatrix} 2.0346 \\ 2.0346 \\ 2.3612 \\ 2.3612 \\ 2.3467 \\ 2.3467 \\ 2.9453 \\ 2.9453 \\ 2.3273 \\ 2.3273 \\ 4.7785 \\ 4.7785 \end{bmatrix} \right\}$$

$$\Omega_s = \left\{ x \in \mathbb{R}^2 : \begin{bmatrix} -0.9996 & -0.0273 \\ 0.9996 & 0.0273 \\ -0.9993 & -0.0369 \\ 0.9993 & 0.0369 \\ -0.9731 & -0.2305 \\ 0.9731 & 0.2305 \\ 0.9164 & 0.4004 \\ -0.9164 & -0.4004 \\ 0.8434 & 0.5372 \\ -0.8434 & -0.5372 \\ 0.7669 & 0.6418 \\ -0.7669 & -0.6418 \\ 0.6942 & 0.7198 \\ -0.6942 & -0.7198 \\ 0.6287 & 0.7776 \\ -0.6287 & -0.7776 \\ 0.5712 & 0.8208 \\ -0.5712 & -0.8208 \end{bmatrix} x \leq \begin{bmatrix} 3.5340 \\ 3.5340 \\ 3.5104 \\ 3.5104 \\ 3.4720 \\ 3.4720 \\ 3.5953 \\ 3.5953 \\ 3.8621 \\ 3.8621 \\ 4.2441 \\ 4.2441 \\ 4.7132 \\ 4.7132 \\ 5.2465 \\ 5.2465 \\ 5.8267 \\ 5.8267 \end{bmatrix} \right\}$$

Figure 2.14 presents state trajectories of the closed loop system with $u(k) = \text{sat}(Kx(k))$ for different initial conditions and realizations of $\alpha(k)$.

Chapter 3
Optimal and Constrained Control—An Overview

3.1 Dynamic Programming

Dynamic programming was developed by R.E. Bellman in the early fifties [13–16]. It provides a sufficient condition for optimality of the control problems for various classes of systems, e.g. linear, time-varying or nonlinear. In general the optimal solution is expressed as a time-varying state-feedback form.

Dynamic programming is based on the following principle of optimality [17],

An optimal policy has the property that whatever the initial state and initial decision are, the remaining decisions must constitute an optimal policy with regard to the state resulting from the first decision.

To begin, let us consider the following optimal control problem,

$$\min_{x,u}\left\{ E\big(x(N)\big) + \sum_{k=0}^{N-1} L\big(x(k), u(k)\big) \right\} \tag{3.1}$$

subject to

$$\begin{cases} x(k+1) = f\big(x(k), u(k)\big), & k = 0, 1, \ldots, N-1, \\ u(k) \in U, & k = 0, 1, \ldots, N-1, \\ x(k) \in X, & k = 0, 1, \ldots, N, \\ x(0) = x_0 \end{cases}$$

where

- $x(k) \in \mathbb{R}^n$ and $u(k) \in \mathbb{R}^m$ are respectively, the state and control variables.
- $N > 0$ is called *the time horizon*.
- $L(x(k), u(k))$ represents a cost along the trajectory.
- $E(x(N))$ represents the terminal cost.
- U and X are respectively, the input and state constraints.
- $x(0)$ is the initial condition.

H.-N. Nguyen, *Constrained Control of Uncertain, Time-Varying, Discrete-Time Systems*,   43
Lecture Notes in Control and Information Sciences 451,
DOI 10.1007/978-3-319-02827-9_3,
© Springer International Publishing Switzerland 2014

Define the *value function* $V_i(x(i))$ as,

$$V_i(x(i)) = \min_{x,u}\left\{ E(x(N)) + \sum_{k=i}^{N-1} L(x(k), u(k)) \right\} \qquad (3.2)$$

subject to

$$\begin{cases} x(k+1) = f(x(k), u(k)), & k = i, i+1, \ldots, N-1, \\ u(k) \in U, & k = i, i+1, \ldots, N-1, \\ x(k) \in X, & k = i, i+1, \ldots, N \end{cases}$$

for $i = N, N-1, N-2, \ldots, 0$.

$V_i(x(i))$ is the optimal cost on the horizon $[i, N]$, starting from the state $x(i)$. Using the principle of optimality, one has,

$$V_i(x(i)) = \min_{u(i)}\left\{ L(x(i), u(i)) + V_{i+1}(x(i+1)) \right\} \qquad (3.3)$$

By substituting $x(i+1) = f(x(i), u(i))$ in (3.3), one obtains,

$$V_i(z) = \min_{u(i)}\left\{ L(x(i), u(i)) + V_{i+1}(f(x(i), u(i))) \right\} \qquad (3.4)$$

subject to

$$\begin{cases} u(i) \in U, \\ f(x(i), u(i)) \in X \end{cases}$$

Problem (3.4) is much simpler than (3.1) because it involves only *one* decision variable $u(i)$. To actually solve this problem, we work backwards in time from $i = N$, starting with

$$V_N(x(N)) = E(x(N))$$

Based on the value function $V_{i+1}(x(i+1))$ with $i = N-1, N-2, \ldots, 0$, the optimal control values $u^*(i)$ can be obtained as,

$$u^*(i) = \arg\min_{u(i)}\left\{ L(x(i), u(i)) + V_{i+1}(f(x(i), u(i))) \right\}$$

subject to

$$\begin{cases} u(i) \in U, \\ f(x(i), u(i)) \in X \end{cases}$$

3.2 Pontryagin's Maximum Principle

The second milestone in the optimal control theory is the Pontryagin's maximum principle [36, 103]. This approach, can be seen as a counterpart of the classical calculus of variation approach, allowing us to solve the control problems in which the control input is subject to constraints in a very general way. Here for illustration, we consider the following simple optimal control problem,

$$\min_{x,u} \left\{ E\big(x(N)\big) + \sum_{k=0}^{N-1} L\big(x(k), u(k)\big) \right\} \tag{3.5}$$

subject to

$$\begin{cases} x(k+1) = f\big(x(k), u(k)\big), & k = 0, 1, \ldots, N-1, \\ u(k) \in U, & k = 0, 1, \ldots, N-1, \\ x(0) = x_0 \end{cases}$$

For simplicity, the state variables are considered unconstrained. For solving the problem (3.5) with the Pontryagin's maximum principle, the following Hamiltonian $H_k(\cdot)$ is defined,

$$H_k\big(x(k), u(k), \lambda(k+1)\big) = L\big(x(k), u(k)\big) + \lambda^T(k+1) f\big(x(k), u(k)\big) \tag{3.6}$$

where $\lambda(k) \in \mathbb{R}^n$ with $k = 1, 2, \ldots, N$ are called the *co-state* or the *adjoint* variables. For the problem (3.5), these variables must satisfy the so called co-state equation,

$$\lambda^*(k+1) = \frac{\partial H_k}{\partial (x(k))}, \quad k = 0, 1, \ldots, N-2$$

and

$$\lambda^*(N) = \frac{\partial E(x(N))}{\partial (x(N))}$$

For given state and co-state variables, the Pontryagin's maximum principle states that the optimal control value is achieved by choosing control $u^*(k)$ that minimizes the Hamiltonian at each time instant, i.e.

$$H_k\big(x^*(k), u^*(k), \lambda^*(k+1)\big) \le H_k\big(x^*(k), u(k), \lambda^*(k+1)\big), \quad \forall u(k) \in U$$

3.3 Model Predictive Control

Model predictive control (MPC), or receding horizon control, is one of the most advanced control approaches which, in the last decades, has became a leading industrial control technology for constrained control systems [30, 34, 47, 52, 88, 92, 107]. MPC is an optimization based strategy, where a model of the plant is used to predict the future evolution of the system, see [88, 92]. This prediction uses the current state of the plant as the initial state and, at each time instant, k, the controller computes a finite optimal control sequence. Then the first control action in this sequence is applied to the plant at time instant k, and at time instant $k+1$ the optimization procedure is repeated with a new plant measurement. This open loop optimal feedback mechanism[1] of the MPC compensates for the prediction error due to structural

---

[1] It was named OLOF (Open Loop Optimal Feedback) control, by the author of [40].

mismatch between the model and the real system as well as for disturbances and measurement noise. In contrast to the maximal principle solutions, which are almost always open loop optimal, the receding horizon principle behind MPC brings the advantage of the feedback structure.

The main advantage which makes MPC industrially desirable is that it can take into account constraints in the control problem. This feature is very important for several reasons,

- Often the best performance, which may correspond to the most efficient operation, is obtained when the system is made to operate near the constraints.
- The possibility to explicitly express constraints in the problem formulation offers a natural way to state complex control objectives.
- Stability and other features can be proved, at least in some cases, in contrast to popular ad-hoc methods to handle constraints, like anti-windup control [50], and override control [118].

3.3.1 Implicit Model Predictive Control

Consider the problem of regulating to the origin the following time-invariant linear discrete-time system,

$$x(k+1) = Ax(k) + Bu(k) \tag{3.7}$$

where $x(k) \in \mathbb{R}^n$ and $u(k) \in \mathbb{R}^m$ are respectively the state and the input variables, $A \in \mathbb{R}^{n \times n}$ and $B \in \mathbb{R}^{n \times m}$. Both $x(k)$ and $u(k)$ are subject to polytopic constraints,

$$\begin{cases} x(k) \in X, & X = \left\{ x \in \mathbb{R}^n : F_x x \leq g_x \right\} \\ u(k) \in U, & U = \left\{ u \in \mathbb{R}^m : F_u u \leq g_u \right\} \end{cases} \quad \forall k \geq 0 \tag{3.8}$$

where the matrices F_x, F_u and the vectors g_x, g_u are assumed to be constant with $g_x > 0$, $g_u > 0$. Here the inequalities are element-wise.

Provided that $x(k)$ is available, the MPC optimization problem is defined as,

$$V\big(x(k)\big) = \min_{\mathbf{u}=[u_0, u_1, \ldots, u_{N-1}]} \left\{ \sum_{t=1}^{N} x_t^T Q x_t + \sum_{t=0}^{N-1} u_t^T R u_t \right\} \tag{3.9}$$

subject to

$$\begin{cases} x_{t+1} = Ax_t + Bu_t, & t = 0, 1, \ldots, N-1, \\ x_t \in X, & t = 1, 2, \ldots, N, \\ u_t \in U, & t = 0, 1, \ldots, N-1, \\ x_0 = x(k) \end{cases}$$

where

- x_{t+1} and u_t are, respectively the predicted states and the predicted inputs, $t = 0, 1, \ldots, N-1$.
- $Q \in \mathbb{R}^{n \times n}$ and $Q \succeq 0$.

- $R \in \mathbb{R}^{m \times m}$ and $R \succ 0$.
- $N \geq 1$ is a fixed integer. N is called *the time horizon* or *the prediction horizon*.

The conditions on Q and R guarantee that the cost function (3.9) is strongly convex. In term of eigenvalues, the eigenvalues of Q should be non-negative, while those of R should be positive in order to ensure the unique optimal solution.

Clearly, the term $x_t^T Q x_t$ penalizes the deviation of the state x from the origin, while the term $u_t^T R u_t$ measures the input control energy. In other words, selecting Q large means that, to keep V small, the state x_t must be as close as possible to the origin in a weighted Euclidean norm. On the other hand, selecting R large means that the control input u_t must be small to keep the cost function V small.

An alternative is a performance measure based on l_1-norm,

$$\min_{\mathbf{u}=[u_0, u_1, \ldots, u_{N-1}]} \left\{ \sum_{t=1}^{N} |Qx_t|_1 + \sum_{t=0}^{N-1} |Ru_t|_1 \right\} \tag{3.10}$$

or l_∞-norm,

$$\min_{\mathbf{u}=[u_0, u_1, \ldots, u_{N-1}]} \left\{ \sum_{t=1}^{N} |Qx_t|_\infty + \sum_{t=0}^{N-1} |Ru_t|_\infty \right\} \tag{3.11}$$

Using the state space model (3.7), the future state variables are expressed sequentially using the set of future control variable values,

$$\begin{cases} x_1 = Ax_0 + Bu_0 \\ x_2 = Ax_1 + Bu_1 = A^2 x_0 + ABu_0 + Bu_1 \\ \vdots \\ x_N = A^N x_0 + A^{N-1} Bu_0 + A^{N-2} Bu_1 + \cdots + Bu_{N-1} \end{cases} \tag{3.12}$$

The set of (3.12) can be rewritten in a compact matrix form as,

$$\mathbf{x} = A_a x_0 + B_a \mathbf{u} = A_a x(k) + B_a \mathbf{u} \tag{3.13}$$

with $\mathbf{x} = [x_1^T\ x_2^T\ \ldots\ x_N^T]^T$, $\mathbf{u} = [u_0^T\ u_1^T\ \ldots\ u_{N-1}^T]^T$ and

$$A_a = \begin{bmatrix} A \\ A^2 \\ \vdots \\ A^N \end{bmatrix}, \qquad B_a = \begin{bmatrix} B & 0 & \cdots & 0 \\ AB & B & \ddots & \vdots \\ \vdots & \vdots & \ddots & 0 \\ A^{N-1}B & A^{N-2}B & \cdots & B \end{bmatrix}$$

The MPC optimization problem (3.9) can be rewritten as,

$$V(x(k)) = \min_{\mathbf{u}} \left\{ \mathbf{x}^T Q_a \mathbf{x} + \mathbf{u}^T R_a \mathbf{u} \right\} \tag{3.14}$$

where

$$Q_a = \begin{bmatrix} Q & 0 & \cdots & 0 \\ 0 & Q & \cdots & 0 \\ \vdots & \vdots & \ddots & \vdots \\ 0 & 0 & \cdots & Q \end{bmatrix}, \qquad R_a = \begin{bmatrix} R & 0 & \cdots & 0 \\ 0 & R & \cdots & 0 \\ \vdots & \vdots & \ddots & \vdots \\ 0 & 0 & \cdots & R \end{bmatrix}$$

and by substituting (3.13) in (3.14), one gets

$$V\big(x(k)\big) = \min_{\mathbf{u}}\{\mathbf{u}^T H\mathbf{u} + 2x^T(k)F\mathbf{u} + x^T(k)Yx(k)\} \qquad (3.15)$$

where

$$H = B_a^T Q_a B_a + R_a, \qquad F = A_a^T Q_a B_a, \qquad Y = A_a^T Q_a A_a \qquad (3.16)$$

Consider now the constraints (3.8) along the horizon. Using (3.8), it can be shown that the constraints on the predicted states and inputs are,

$$\begin{cases} F_x^a \mathbf{x} \le g_x^a, \\ F_u^a \mathbf{u} \le g_u^a \end{cases} \qquad (3.17)$$

where

$$F_x^a = \begin{bmatrix} F_x & 0 & \cdots & 0 \\ 0 & F_x & \cdots & 0 \\ \vdots & \vdots & \ddots & \vdots \\ 0 & 0 & \cdots & F_x \end{bmatrix}, \qquad g_x^a = \begin{bmatrix} g_x \\ g_x \\ \vdots \\ g_x \end{bmatrix},$$

$$F_u^a = \begin{bmatrix} F_u & 0 & \cdots & 0 \\ 0 & F_u & \cdots & 0 \\ \vdots & \vdots & \ddots & \vdots \\ 0 & 0 & \cdots & F_u \end{bmatrix}, \qquad g_u^a = \begin{bmatrix} g_u \\ g_u \\ \vdots \\ g_u \end{bmatrix}$$

Using (3.13), the state constraints along the horizon can be expressed as,

$$F_x^a\{A_a x(k) + B_a\mathbf{u}\} \le g_x^a$$

or, equivalently

$$F_x^a B_a \mathbf{u} \le -F_x^a A_a x(k) + g_x^a \qquad (3.18)$$

Combining (3.17), (3.18), one obtains,

$$G\mathbf{u} \le Ex(k) + S \qquad (3.19)$$

where

$$G = \begin{bmatrix} F_u^a \\ F_x^a B_a \end{bmatrix}, \qquad E = \begin{bmatrix} 0 \\ -F_x^a A_a \end{bmatrix}, \qquad S = \begin{bmatrix} g_u^a \\ g_x^a \end{bmatrix}$$

From (3.14) and (3.19), the MPC problem can be formulated as,

$$V_1\big(x(k)\big) = \min_{\mathbf{u}}\{\mathbf{u}^T H\mathbf{u} + 2x^T(k)F\mathbf{u}\} \qquad (3.20)$$

subject to

$$G\mathbf{u} \le Ex(k) + S$$

where the term $x^T(k)Yx(k)$ is removed since it does not influence the optimal argument. The value of the cost function at optimum is simply obtained from (3.20) by,

$$V\big(x(k)\big) = V_1\big(x(k)\big) + x^T(k)Yx(k)$$

The control law obtained by solving on-line (3.20) is called *implicit* model predictive control.

Example 3.1 Consider the following time-invariant linear discrete-time system,

$$x(k+1) = \begin{bmatrix} 1 & 1 \\ 0 & 1 \end{bmatrix} x(k) + \begin{bmatrix} 1 \\ 0.7 \end{bmatrix} u(k) \tag{3.21}$$

and the MPC problem with $N = 3$, $Q = I$ and $R = 1$. The constraints are,

$$-2 \le x_1 \le 2, \qquad -5 \le x_2 \le 5, \qquad -1 \le u \le 1$$

Using (3.20), the MPC problem can be described as,

$$\min_{\mathbf{u} = \{u_0, u_1, u_2\}} \left\{ \mathbf{u}^T H \mathbf{u} + 2x^T(k) F \mathbf{u} \right\}$$

where

$$H = \begin{bmatrix} 12.1200 & 6.7600 & 2.8900 \\ 6.7600 & 5.8700 & 2.1900 \\ 2.8900 & 2.1900 & 2.4900 \end{bmatrix}, \qquad F = \begin{bmatrix} 5.1000 & 2.7000 & 1.0000 \\ 13.7000 & 8.5000 & 3.7000 \end{bmatrix}$$

and subject to

$$G\mathbf{u} \le S + Ex(k)$$

where

$$G = \begin{bmatrix} 1.0000 & 0 & 0 \\ -1.0000 & 0 & 0 \\ 0 & 1.0000 & 0 \\ 0 & -1.0000 & 0 \\ 0 & 0 & 1.0000 \\ 0 & 0 & -1.0000 \\ 1.0000 & 0 & 0 \\ 0.7000 & 0 & 0 \\ -1.0000 & 0 & 0 \\ -0.7000 & 0 & 0 \\ 1.7000 & 1.0000 & 0 \\ 0.7000 & 0.7000 & 0 \\ -1.7000 & -1.0000 & 0 \\ -0.7000 & -0.7000 & 0 \\ 2.4000 & 1.7000 & 1.0000 \\ 0.7000 & 0.7000 & 0.7000 \\ -2.4000 & -1.7000 & -1.0000 \\ -0.7000 & -0.7000 & -0.7000 \end{bmatrix}, \quad E = \begin{bmatrix} 0 & 0 \\ 0 & 0 \\ 0 & 0 \\ 0 & 0 \\ 0 & 0 \\ 0 & 0 \\ -1 & -1 \\ 0 & -1 \\ 1 & 1 \\ 0 & 1 \\ -1 & -2 \\ 0 & -1 \\ 1 & 2 \\ 0 & 1 \\ -1 & -3 \\ 0 & -1 \\ 1 & 3 \\ 0 & 1 \end{bmatrix}, \quad S = \begin{bmatrix} 1 \\ 1 \\ 1 \\ 1 \\ 1 \\ 1 \\ 2 \\ 5 \\ 2 \\ 5 \\ 2 \\ 5 \\ 2 \\ 5 \\ 2 \\ 5 \\ 2 \\ 5 \end{bmatrix}$$

Algorithm 3.1 Model predictive control—Implicit approach

1. Measure the current state $x(k)$.
2. Compute the control signal sequence $\mathbf{u}$ by solving (3.20).
3. Apply first element of $\mathbf{u}$ as input to the system (3.7).
4. Wait for the next time instant $k := k + 1$.
5. Go to step 1 and repeat

Fig. 3.1 State and input
trajectories as functions of
time for Example 3.1

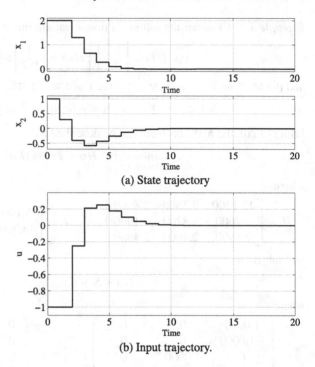

(a) State trajectory

(b) Input trajectory.

For the initial condition $x(0) = [2\ 1]^T$, Fig. 3.1 shows the state and input trajectories
of the closed loop system obtained by using Algorithm 3.1.

3.3.2 Recursive Feasibility and Stability

Recursive feasibility of the optimization problem and stability of the resulting
closed-loop system are two important aspects when designing a MPC controller.

Recursive feasibility of the QP problem (3.20) means that if (3.20) is feasible at
time k, it will be also feasible at time $k + 1$. In other words there exists an admissible
control value that holds the system within the state constraints. The feasibility problem can arise due to model errors, disturbances or the choice of the cost function.

Stability analysis necessitates the use of Lyapunov theory [70], since the presence of the constraints makes the closed-loop system nonlinear. In addition, it
is well known that unstable input-constrained system cannot be globally stabi-
lized [89, 113, 119]. Another problem is that the control law is generated by the
solution of the QP problem (3.20) and generally there does not exist any simple
closed-form expression for the solution, although it can be shown that the solution
is a piecewise affine state feedback law [20].

Recursive feasibility and stability can be assured by adding a terminal cost function in the objective function (3.9) and by including the final state of the planning

horizon in a terminal invariant set. Let the matrix $P \in \mathbb{R}^{n \times n}$ be the unique solution of the following discrete-time algebraic Riccati equation,

$$P = A^T P A - A^T P B (B^T X B + R)^{-1} B^T P A + Q \qquad (3.22)$$

and the matrix gain $K \in \mathbb{R}^{m \times n}$ is defined as,

$$K = -(B^T P B + R)^{-1} B^T P A \qquad (3.23)$$

It is well known [7, 80, 83, 85] that matrix gain K is the solution of the optimization problem (3.9) when the time horizon $N = \infty$ and there are no active state and input constraints. In this case the cost function is,

$$V(x(0)) = \sum_{k=0}^{\infty} \{x_k^T Q x_k + u_k^T R u_k\} = \sum_{k=0}^{\infty} x_k^T (Q + K^T R K) x_k = x_0^T P x_0$$

For the stabilizing controller $u(k) = K x(k)$, using results in Sect. 2.3, an ellipsoidal or polyhedral invariant set $\Omega \subseteq X$ can be computed for system (3.7) with constraints (3.8).

Consider now the following MPC optimization problem,

$$\min_{\mathbf{u} = [u_0, u_1, \dots, u_{N-1}]} \left\{ x_N^T P x_N + \sum_{t=0}^{N-1} \{x_t^T Q x_t + u_t^T R u_t\} \right\} \qquad (3.24)$$

subject to

$$\begin{cases} x_{t+1} = A x_t + B u_t, & t = 0, 1, \dots, N-1, \\ x_t \in X, & t = 1, 2, \dots, N-1, \\ u_t \in U, & t = 0, 1, \dots, N-1, \\ x_N \in \Omega, \\ x_0 = x(k) \end{cases}$$

The following theorem holds [92]

Theorem 3.1 [92] *Assuming feasibility at the initial state, the MPC controller (3.24) guarantees recursive feasibility and asymptotic stability.*

The MPC problem considered here uses both a terminal cost function and a terminal set constraint and is called the dual-mode MPC. This MPC scheme is the most attractive version in the MPC literature. In general, it offers better performance compared with other MPC versions and allows a wider range of control problems to be handled. The downside is the dependence of the feasible domain on the prediction horizon. Generally, for a large domain one needs to employ a large prediction horizon.

3.3.3 Explicit Model Predictive Control—Parameterized Vertices

Note that the implicit MPC requires running on-line optimization algorithms to solve a QP problem associated with the objective function (3.9) or to solve an LP

problem with the objective function (3.10) or (3.11). Although computational speed
and optimization algorithms are continuously improving, solving a QP or LP problem can be computationally costly, specially when the prediction horizon is large,
and this has traditionally limited MPC to applications with relatively low complex-
ity/sampling interval ratio.

Indeed the state vector can be interpreted as a vector of parameters in the optimization problem (3.24). The exact optimal solution can be expressed as *a piecewise affine function* of the state over a polyhedral partition of the state space and
the MPC computation can be moved off-line [20, 30, 95, 115]. The control action is
then computed on-line by lookup tables and search trees.

Several solutions have been proposed in the literature for constructing a polyhedral partition of the state space [20, 95, 115]. In [19, 20] some iterative techniques
use a QP or LP to find feasible points and then split the parameters space by in-
verting one by one the constraints hyper-planes. As an alternative, in [115] the au-
thors construct the unconstrained polyhedral region and then enumerate the others
based on the combinations of active constraints. When the cost function is quadratic,
the uniqueness of the optimal solution is guaranteed and the methods proposed in
[19, 20, 115] work well, at least for non-degenerate sets of constraints [127].

It is worth noticing that by using l_1- or l_∞-norms as the performance measure,
the cost function is only positive semi-definite and the uniqueness of the optimal
solution is not guaranteed and as a consequence, neither the continuity. A control
law will have a practical advantage if the control action presents no jumps on the
boundaries of the polyhedral partitions. When the optimal solution is not unique,
the methods in [19, 20, 115] allow discontinuities as long as during the exploration
of the parameters space, the optimal basis is chosen arbitrarily.

Note that using the cost (3.10) or (3.11), the MPC problem can be rewritten as

$$V\left(x(k)\right) = \min_z \{c^T z\} \tag{3.25}$$

subject to

$$G_l z \le E_l x(k) + S_l$$

with

$$z = \begin{bmatrix} \xi_1^T & \xi_2^T & \cdots & \xi_{N_\xi}^T & u_0^T & u_1^T & \cdots & u_{N-1}^T \end{bmatrix}^T$$

where $\xi_i, i = 1, 2, \ldots, N_\xi$ are slack variables and N_ξ depends on the norm used and
on the prediction horizon N. Details of how to compute vectors c, S_l and matrices
G_l, E_l are well known [19].

The feasible domain for the LP problem (3.25) is defined by a finite number of
inequalities with a right hand side linearly dependent on the vector of parameters
$x(k)$, describing in fact a *parameterized polytope* [86],

$$P\left(x(k)\right) = \{z : G_l z \le E_l x(k) + S_l\} \tag{3.26}$$

For simplicity, it is assumed that, the polyhedral set $P(x(k))$ is bounded
$\forall x(k) \in X$. In this case $P(x(k))$ can be expressed in a dual (generator based) form
as

$$P(x) = \text{Conv}\{v_i\left(x(k)\right)\}, \quad i = 1, 2, \ldots, n_v \tag{3.27}$$

where $v_i(x(k))$ are the parameterized vertices. Each parameterized vertex in (3.27) is characterized by a set of active constraints. Once the set of active constraints is identified, the parameterized vertex $v_i(x(k))$ can be computed as

$$v_i\left(x(k)\right) = \overline{G}_{li}^{-1}\overline{E}_{li}x(k) + \overline{G}_{li}^{-1}\overline{S}_{li} \tag{3.28}$$

where $\overline{G}_{li}, \overline{E}_{li}, \overline{W}_{li}$ correspond to the subset of active constraints for the i-th parameterized vertex.

As a first conclusion, the construction of the dual description (3.26), (3.27) requires the determination of the set of parameterized vertices. Efficient algorithms exist in this direction [86], the main idea is the analogy with a non-parameterized polytope in a higher dimension.

When the vector of parameter $x(k)$ varies inside X, the vertices (3.27) may split or merge. This means that each parameterized vertex $v_i(x(k))$ is defined only over a specific region in the parameters space. These regions VD_i are called *validity domains* and can be constructed using simple projection mechanism [86].

Once the entire family of parameterized vertices and their validity domains are available, the optimal solution can be constructed as follows.

For a given $x(k)$, the minimum (3.25) is attained by a subset of vertices $v_i(x(k))$ of $P(x(k))$, denoted $v_i^*(x(k))$. The complete solution is,

$$z_k\left(x(k)\right) = \text{Conv}\left\{v_{1k}^*\left(x(k)\right), v_{2k}^*\left(x(k)\right), \dots, v_{sk}^*\left(x(k)\right)\right\} \tag{3.29}$$

i.e. any function included in the convex combination of vertices is optimal.

The following theorem holds regarding the structure of the polyhedral partitions of the parameters space [96].

Theorem 3.2 [96] *Let the multi-parametric program in (3.25) and $v_i(x(k))$ be the parameterized vertices of the feasible domain (3.26), (3.27) with their corresponding validity domains VD_i. If a parameterized vertex is selected as an optimal candidate, then it covers all its validity domain.*

It is worth noticing that the complete optimal solution (3.29) takes into account the eventual non-uniqueness of the optimum, and it defines the entire family of optimal solutions using the parameterized vertices and their validity domains.

Once the entire family of optimal solutions is available, the continuity of the control law can be guaranteed as follows. First if the optimal solution is unique, then there is no decision to be made, the explicit solution being the collection of the parameterized vertices and their validity domains. The continuity is intrinsic.

Conversely, the family of the optimal solutions can be enriched in the presence of several optimal parameterized vertices,

$$\begin{cases} z_k\left(x(k)\right) = \alpha_{1k}v_{1k}^* + \alpha_{2k}v_{2k}^* + \cdots + \alpha_{sk}v_{sk}^* \\ \alpha_{ik} \geq 0, \quad i = 1, 2, \dots, s \\ \alpha_{1k} + \alpha_{2k} + \cdots + \alpha_{sk} = 1 \end{cases} \tag{3.30}$$

passing to an infinite number of candidates. As mentioned previously, the vertices of the feasible domain split and merge. The changes occur with a preservation of the continuity. Hence the continuity of the control law is guaranteed by the continuity of the parameterized vertices. The interested reader is referred to [96] for further discussions on the related concepts and constructive procedures.

Example 3.2 To illustrate the parameterized vertices concept, consider the following feasible domain for the MPC optimization problem,

$$P(x(k)) = P_1 \cap P_2(x(k)) \tag{3.31}$$

where P_1 is a fixed polytope,

$$P_1 = \left\{ z \in \mathbb{R}^2 : \begin{bmatrix} 0 & 1 \\ 1 & 0 \\ 0 & -1 \\ -1 & 0 \end{bmatrix} z \leq \begin{bmatrix} 1 \\ 1 \\ 0 \\ 0 \end{bmatrix} \right\} \tag{3.32}$$

and $P_2(x(k))$ is a parameterized polyhedral set,

$$P_2(x(k)) = \left\{ z \in \mathbb{R}^2 : \begin{bmatrix} -1 & 0 \\ 0 & -1 \end{bmatrix} z \leq \begin{bmatrix} -1 \\ -1 \end{bmatrix} x(k) + \begin{bmatrix} 0.5 \\ 0.5 \end{bmatrix} \right\} \tag{3.33}$$

Note that $P_2(x(k))$ is an unbounded set. Using (3.33), it is follows that,

- If $x(k) \leq 0.5$, then $-x(k) + 0.5 \geq 0$. It follows that $P_1 \subset P_2(x(k))$, and hence $P(x(k)) = P_1$ has the half-space representation as (3.32) and the vertex representation as,

$$P(x(k)) = \text{Conv}\{v_1, v_2, v_3, v_4\}$$

where

$$v_1 = \begin{bmatrix} 0 \\ 0 \end{bmatrix}, \qquad v_2 = \begin{bmatrix} 1 \\ 0 \end{bmatrix}, \qquad v_3 = \begin{bmatrix} 0 \\ 1 \end{bmatrix}, \qquad v_4 = \begin{bmatrix} 1 \\ 1 \end{bmatrix}$$

- If $0.5 \leq x(k) \leq 1.5$, then $-1 \leq -x(k) + 0.5 \leq 0$. It follows that $P_1 \cap P_2(x(k)) \neq \emptyset$. Note that for $P(x(k))$, the half-spaces $z_1 = 0$ and $z_2 = 0$ are redundant. $P(x(k))$ has the half-space representation,

$$P(x(k)) = \left\{ z \in \mathbb{R}^2 : \begin{bmatrix} 0 & 1 \\ 1 & 0 \\ -1 & 0 \\ 0 & -1 \end{bmatrix} z \leq \begin{bmatrix} 0 \\ 0 \\ -1 \\ -1 \end{bmatrix} x(k) + \begin{bmatrix} 1 \\ 1 \\ 0.5 \\ 0.5 \end{bmatrix} \right\}$$

and the vertex representation,

$$P(x(k)) = \text{Conv}\{v_4, v_5, v_6, v_7\}$$

with

$$v_5 = \begin{bmatrix} 1 \\ x - 0.5 \end{bmatrix}, \qquad v_6 = \begin{bmatrix} x - 0.5 \\ 1 \end{bmatrix}, \qquad v_7 = \begin{bmatrix} x - 0.5 \\ x - 0.5 \end{bmatrix},$$

- If $1.5 < x(k)$, then $-x(k) + 0.5 < -1$. It follows that $P_1 \cap P_2(x(k)) = \emptyset$. Hence $P(x(k)) = \emptyset$.

Table 3.1 Validity domains and their parameterized vertices

	VD_1	VD_2	VD_3
	v_1, v_2, v_3, v_4	v_4, v_5, v_6, v_7	$\emptyset$

Fig. 3.2 Polyhedral sets P_1 and $P_2(x(k))$ with $x(k) = 0.3$, $x(k) = 0.9$ and $x(k) = 1.5$ for Example 3.2. For $x(k) \leq 0.5$, $P_1 \cap P_2(x(k)) = P_1$. For $0.5 \leq x(k) \leq 1.5$, $P_1 \cap P_2(x(k)) \neq \emptyset$. For $x(k) > 1.5$, $P_1 \cap P_2(x(k)) = \emptyset$

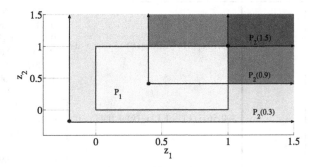

In conclusion, the parameterized vertices of $P(x(k))$ are,

$$v_1 = \begin{bmatrix} 0 \\ 0 \end{bmatrix}, \qquad v_2 = \begin{bmatrix} 1 \\ 0 \end{bmatrix}, \qquad v_3 = \begin{bmatrix} 0 \\ 1 \end{bmatrix}, \qquad v_4 = \begin{bmatrix} 1 \\ 1 \end{bmatrix},$$

$$v_5 = \begin{bmatrix} 1 \\ x(k) - 0.5 \end{bmatrix}, \qquad v_6 = \begin{bmatrix} x(k) - 0.5 \\ 1 \end{bmatrix}, \qquad v_7 = \begin{bmatrix} x(k) - 0.5 \\ x(k) - 0.5 \end{bmatrix},$$

and the validity domains,

$$VD_1 = (-\infty \quad 0.5], \qquad VD_2 = [0.5 \quad 1.5], \qquad VD_3 = (1.5 \quad +\infty)$$

Table 3.1 presents the validity domains and their corresponding parameterized vertices.

Figure 3.2 shows the polyhedral sets P_1 and $P_2(x(k))$ with $x(k) = 0.3$, $x(k) = 0.9$ and $x(k) = 1.5$.

Example 3.3 Consider the linear discrete-time system in Example 3.1 with the same constraints on the state and input variables. Here we will use a dual-mode MPC (3.24), which guarantees recursive feasibility and stability.

By solving (3.22) and (3.23) for $Q = I$, $R = 1$, one obtains,

$$P = \begin{bmatrix} 1.5076 & -0.1173 \\ -0.1173 & 1.2014 \end{bmatrix}, \qquad K = [-0.7015 \quad -1.0576]$$

For the controller $u = Kx$, the terminal invariant set Ω is computed using Procedure 2.2,

$$\Omega = \left\{ x \in \mathbb{R}^2 : \begin{bmatrix} 0.7979 & -0.6029 \\ -0.7979 & 0.6029 \\ 1.0000 & 0 \\ -1.0000 & 0 \\ -0.5528 & -0.8333 \\ 0.5528 & 0.8333 \end{bmatrix} x \leq \begin{bmatrix} 2.5740 \\ 2.5740 \\ 2.0000 \\ 2.0000 \\ 0.7879 \\ 0.7879 \end{bmatrix} \right\}$$

Figure 3.3 shows the state space partition obtained by using the parameterized vertices framework as a method to construct the explicit solution to the MPC problem (3.24) with prediction horizon $N = 2$.

The control law over the state space partition is,

$$
u(k) = \begin{cases}
-0.70x_1(k) - 1.06x_2(k) & \text{if} \quad \begin{bmatrix} -0.80 & 0.60 \\ 0.80 & -0.60 \\ 0.55 & 0.83 \\ -0.55 & -0.83 \\ -1.00 & 0.00 \\ 1.00 & 0.00 \end{bmatrix} x(k) \leq \begin{bmatrix} 2.57 \\ 2.57 \\ 0.79 \\ 0.79 \\ 2.00 \\ 2.00 \end{bmatrix} \\
& \qquad \text{(Region 1)} \\[2pt]
-0.56x_1(k) - 1.17x_2(k) + 0.47 & \text{if} \quad \begin{bmatrix} 0.43 & 0.90 \\ -1.00 & 0.00 \\ 0.80 & -0.60 \end{bmatrix} x(k) \leq \begin{bmatrix} 1.14 \\ 2.00 \\ -2.57 \end{bmatrix} \\
& \qquad \text{(Region 4)} \\[2pt]
-0.56x_1(k) - 1.17x_2(k) - 0.47 & \text{if} \quad \begin{bmatrix} -0.43 & -0.90 \\ 1.00 & 0.00 \\ -0.80 & 0.60 \end{bmatrix} x(k) \leq \begin{bmatrix} 1.14 \\ 2.00 \\ -2.57 \end{bmatrix} \\
& \qquad \text{(Region 7)} \\[2pt]
-1 & \text{if} \quad \begin{bmatrix} 0.37 & 0.93 \\ 1.00 & 0.00 \\ -0.55 & -0.83 \end{bmatrix} x(k) \leq \begin{bmatrix} 1.29 \\ 2.00 \\ -0.79 \end{bmatrix} \\
& \qquad \text{(Region 2)} \\[2pt]
1 & \text{if} \quad \begin{bmatrix} -0.37 & -0.93 \\ -1.00 & 0.00 \\ 0.55 & 0.83 \end{bmatrix} x(k) \leq \begin{bmatrix} 1.29 \\ 2.00 \\ -0.79 \end{bmatrix} \\
& \qquad \text{(Region 5)} \\[2pt]
-1 & \text{if} \quad \begin{bmatrix} 0.71 & 0.71 \\ 0.27 & 0.96 \\ -1.00 & 0.00 \\ 1.00 & 0.00 \\ -0.43 & -0.90 \\ -0.37 & -0.93 \end{bmatrix} x(k) \leq \begin{bmatrix} 2.12 \\ 1.71 \\ 2.00 \\ 2.00 \\ -1.14 \\ -1.29 \end{bmatrix} \\
& \qquad \text{(Region 3)} \\[2pt]
1 & \text{if} \quad \begin{bmatrix} -0.71 & -0.71 \\ -0.27 & -0.96 \\ 1.00 & 0.00 \\ -1.00 & 0.00 \\ 0.43 & 0.90 \\ 0.37 & 0.93 \end{bmatrix} x(k) \leq \begin{bmatrix} 2.12 \\ 1.71 \\ 2.00 \\ 2.00 \\ -1.14 \\ -1.29 \end{bmatrix} \\
& \qquad \text{(Region 6)}
\end{cases}
$$

For the initial condition $x(0) = [-2\ 2.33]$, Fig. 3.4 shows the state and input trajectories as functions of time.

3.4 Vertex Control

The vertex control framework was first proposed by Gutman and Cwikel in [53]. It gives a necessary and sufficient condition for stabilizing a time-invariant linear

Fig. 3.3 State space partition for Example 3.3. Number of regions $N_r = 7$

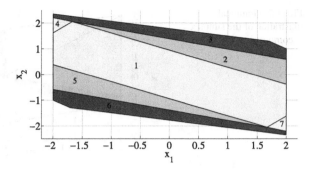

Fig. 3.4 State and input trajectories of the closed loop system as functions of time for Example 3.3

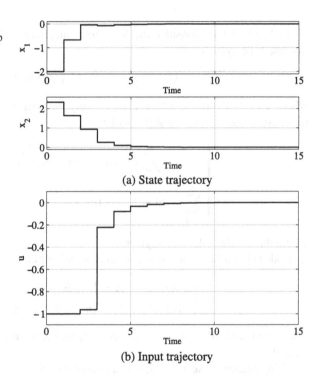

(a) State trajectory

(b) Input trajectory

discrete-time system with bounded polyhedral state and control constraints. The condition is that at each vertex of the controlled invariant set[2] C_N there exists an admissible control action that brings the state to the interior of the set C_N in *finite time*. Then, this condition was extended to the uncertain plant case by Blanchini in [23]. A stabilizing controller is given by the convex combination of vertex controls in each sector with a Lyapunov function given by shrunken images of the boundary of the set C_N [23, 53].

---

[2]See Sect. 2.3.4.

Fig. 3.5 Graphical illustration of the simplex decomposition

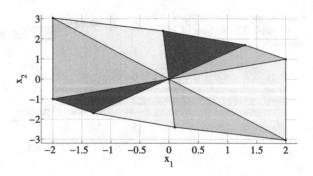

To begin, let us consider the following uncertain and/or time-varying linear discrete-time system,

$$x(k+1) = A(k)x(k) + B(k)u(k) \tag{3.34}$$

where $x(k) \in \mathbb{R}^n$, $u(k) \in \mathbb{R}^m$ are, respectively the state and input vectors. The matrices $A(k) \in \mathbb{R}^{n \times n}$, $B(k) \in \mathbb{R}^{n \times m}$ satisfy,

$$\begin{cases} A(k) = \sum_{i=1}^{q} \alpha_i(k)A_i, \quad B(k) = \sum_{i=1}^{q} \alpha_i(k)B_i, \\ \sum_{i=1}^{q} \alpha_i(k) = 1, \quad \alpha_i(k) \geq 0 \end{cases} \tag{3.35}$$

where the matrices A_i, B_i are given.

Both $x(k)$ and $u(k)$ are subject to the following bounded polytopic constraints,

$$\begin{cases} x(k) \in X, \quad X = \left\{ x \in \mathbb{R}^n : F_x x \leq g_x \right\} \\ u(k) \in U, \quad U = \left\{ u \in \mathbb{R}^m : F_u u \leq g_u \right\} \end{cases} \tag{3.36}$$

where the matrices F_x, F_u and the vectors g_x and g_u are assumed to be constant with $g_x > 0$, $g_u > 0$.

Using results in Sect. 2.3.4, it is assumed that the robustly controlled invariant set C_N with some fixed integer $N > 0$ is available as,

$$C_N = \left\{ x \in \mathbb{R}^n : F_N x \leq g_N \right\} \tag{3.37}$$

The set C_N with non-empty interior can be decomposed as a sequence of simplicies $C_N^{(j)}$ each formed by n vertices $\{x_1^{(j)}, x_2^{(j)}, \ldots, x_n^{(j)}\}$ and the origin, having the following properties, see Fig. 3.5,

- $C_N^{(j)}$ has nonempty interior.
- $\text{Int}(C_N^{(j)}) \cap \text{Int}(C_N^{(l)}) = \emptyset, \forall j \neq l$.
- $\bigcup_j C_N^{(j)} = C_N$.

Denote by $V^{(j)} = [v_1^{(j)} \ v_2^{(j)} \ \dots \ v_n^{(j)}]$ the square matrix defined by the vertices generating $C_N^{(j)}$. Since $C_N^{(j)}$ has nonempty interior, $V^{(j)}$ is invertible. Let

$$U^{(j)} = [u_1^{(j)} \ u_2^{(j)} \ \dots \ u_n^{(j)}]$$

be the $m \times n$ matrix defined by chosen admissible control values[3] satisfying (3.36) at the vertices of $C_N^{(j)}$. Consider the following piecewise linear controller,

$$u(k) = K^{(j)} x(k) \tag{3.38}$$

called the *vertex control law*, for $x(k) \in C_N^{(j)}$, where

$$K^{(j)} = U^{(j)} (V^{(j)})^{-1} \tag{3.39}$$

A simple LP problem is formulated to determine to which simplex j the current state belongs [65, 128] in order to compute (3.38).

Remark 3.1 Generally, one would like to push the state away from the boundary of the set C_N *as far as possible* in a contractive sense: if $x(k)$ is a vertex of C_N, one would like to find the control value $u(k) \in U$, such that $x(k+1) \in \mu C_N$, whereby μ is minimal. This can be done by solving the following linear programming problem,

$$J = \min_{\mu, u} \{\mu\} \tag{3.40}$$

subject to

$$\begin{cases} F_N(A_i x + B_i u) \le \mu g_N, & \forall i = 1, 2 \dots, q, \\ F_u u \le g_u, \\ 0 \le \mu \le 1 \end{cases}$$

Due to the invariance properties of C_N, problem (3.40) is always feasible.

Theorem 3.3 *The vertex controller* (3.38) *guarantees recursive feasibility for all initial states $x(0) \in C_N$.*

Proof Proofs are given in [23, 53]. Here a new simple proof is proposed. For all $x(k) \in C_N$, there exists an index j such that $x(k) \in C_N^{(j)}$, and $x(k)$ can be expressed as a convex combination of the vertices of $C_N^{(j)}$,

$$x(k) = \beta_1(k) v_1^{(j)} + \beta_2(k) v_2^{(j)} + \dots + \beta_n(k) v_n^{(j)} \tag{3.41}$$

where $\sum_{i=1}^{n} \beta_i(k) \le 1$ and $\beta_i(k) \ge 0$. Equation (3.41) can be written in a compact form as,

$$x(k) = V^{(j)} \beta(k) \tag{3.42}$$

---

[3]By an admissible control value we understand any control value that is the first of a sequence of control values that bring the state from the vertex to the interior of the feasible set in a *finite number* of steps, see [53].

where $\beta(k) = [\beta_1(k)\ \beta_2(k)\ \ldots\ \beta_n(k)]^T$ and by consequence

$$\beta(k) = \left(V^{(j)}\right)^{-1} x(k) \tag{3.43}$$

Using (3.38), (3.39), one has

$$u(k) = K^{(j)} x(k) = U^{(j)}\left(V^{(j)}\right)^{-1} x(k) \tag{3.44}$$

and by (3.43), equation (3.44) becomes

$$u(k) = U^{(j)}\beta(k) = \sum_{i=1}^{n} \beta_i(k) u_i^{(j)} \tag{3.45}$$

Hence for a given $x(k) \in C_N^{(j)}$, the control value is the convex combination of vertex controls of $C_N^{(j)}$.

For feasibility, one has to assure,

$$\forall x(k) \in C_N : \quad \begin{cases} F_u u(k) \le g_u, \\ x(k+1) = A(k)x(k) + B(k)u(k) \in C_N \end{cases}$$

For the input constraints, using (3.45), it holds that,

$$F_u u(k) = \sum_{i=1}^{n} \beta_i(k) F_u u_i^{(j)} \le \sum_{i=1}^{n} \beta_i(k) g_u \le g_u$$

For the state constraints, using (3.42), (3.45), one obtains,

$$x(k+1) = A(k)x(k) + B(k)u(k) = A(k)V^{(j)}\beta(k) + B(k)U^{(j)}\beta(k)$$
$$= \sum_{i=1}^{n} \beta_i(k)\left(A(k)v_i^{(j)} + B(k)u_i^{(j)}\right)$$

Since $A(k)v_i^{(j)} + B(k)u_i^{(j)} \in C_N, \forall i = 1, 2, \ldots, n$, it follows that $x(k+1) \in C_N$. $\square$

Theorem 3.4 *Given the system* (3.34) *and the constraints* (3.36), *the vertex control law* (3.38) *is robustly asymptotically stabilizing for all initial states* $x(0) \in C_N$.

Proof Proofs are given in [23, 53]. Here we give another proof providing valuable insight into the vertex control scheme. Denote the vertices of C_N as $\{v_1, v_2, \ldots, v_s\}$ where s is the number of vertices. It follows from [53] that the vertex control law can be obtained by solving the following optimization problem, for a given $x(k) \in C_N$,

$$V\big(x(k)\big) = \min_{\beta_i(k)}\left\{\sum_{i=1}^{s} \beta_i(k)\right\} \tag{3.46}$$

subject to

$$\begin{cases} x(k) = \sum_{i=1}^{s} \beta_i(k)v_i, \\ 0 \le \beta_i(k) \le 1, \quad \forall i = 1, 2, \ldots, s \end{cases}$$

and letting the control value $u(k)$ at the current state $x(k)$ be defined as,

$$u(k) = \sum_{i=1}^{s} \beta_i^*(k) u_i$$

where u_i is the vertex control value at v_i and $\beta_i^*(k)$ is the solution of (3.46).

Consider the positive function $V(x) = \sum_{i=1}^{s} \beta_i^*(x)$ for all $x \in C_N$. $V(x)$ is a Lyapunov function candidate. One has

$$x(k+1) = A(k)x(k) + B(k)u(k) = A(k) \sum_{i=1}^{s} \beta_i^*(k)v_i + B(k) \sum_{i=1}^{s} \beta_i^*(k)u_i$$

$$= \sum_{i=1}^{s} \beta_i^*(k)\big(A(k)v_i + B(k)u_i\big) = \sum_{i=1}^{s} \beta_i^*(k)v_i^+(k)$$

where $v_i^+(k) = A(k)v_i + B(k)u_i \in C_N$. Clearly, $v_i^+(k)$ can be expressed as the convex hull of vertices of C_N, i.e.

$$v_i^+(k) = \sum_{j=1}^{s} \gamma_{ij}(k)v_j$$

where $\sum_{j=1}^{s} \gamma_{ij}(k) \leq 1$ and $0 \leq \gamma_{ij}(k) \leq 1$, $\forall j = 1, 2, \ldots, s$, $\forall i = 1, 2, \ldots, s$. It follows that,

$$x(k+1) = \sum_{i=1}^{s} \beta_i^*(k)v_i^+(k) = \sum_{i=1}^{s} \beta_i^*(k) \sum_{j=1}^{s} \gamma_{ij}(k)v_j = \sum_{j=1}^{s} \left(\sum_{i=1}^{s} \beta_i^*(k)\gamma_{ij}(k) \right) v_j \tag{3.47}$$

Hence $(\sum_{i=1}^{s} \beta_i^*(k)\gamma_{ij}(k))$, $\forall j = 1, 2, \ldots, s$ is a feasible solution of the optimization problem (3.46) at time $k+1$. Since $\sum_{j=1}^{s} \gamma_{ij} \leq 1$, $\forall i = 1, 2, \ldots, s$, it follows that,

$$\sum_{j=1}^{s} \left(\sum_{i=1}^{s} \beta_i^*(k)\gamma_{ij} \right) = \sum_{i=1}^{s} \beta_i^*(k) \left(\sum_{j=1}^{s} \gamma_{ij} \right) \leq \sum_{i=1}^{s} \beta_i^*(k) \tag{3.48}$$

By solving the linear programming problem (3.46) at time $k+1$, one gets the optimal solution, namely

$$x(k+1) = \sum_{i=1}^{s} \beta_i^*(k+1)v_i \tag{3.49}$$

Using (3.47), (3.49), it follows that

$$\sum_{i=1}^{s} \beta_i^*(k+1) \leq \sum_{j=1}^{s} \sum_{i=1}^{s} \beta_i^*(k)\gamma_{ij}(k)$$

and together with (3.48), one obtains

$$\sum_{i=1}^{s} \beta_i^*(k+1) \leq \sum_{i=1}^{s} \beta_i^*(k)$$

hence $V(x)$ is a non-increasing function.

Fig. 3.6 Invariant set C_N and state space partition for vertex control for Example 3.4

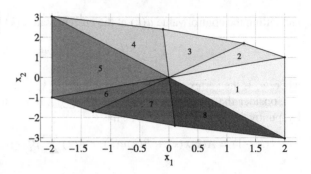

The facts that the level curves of the function $V(x) = \sum_{i=1}^{s} \beta_i^*(k)$ are given by scaling the boundary of C_N, and the state contraction property of applying of the control values u_i at the vertices of C_N, guarantees that there is no initial condition $x(0)$ on the boundary of C_N such that $\sum_{i=1}^{s} \beta_i^*(k) = \sum_{i=1}^{s} \beta_i^*(0) = 1$ for sufficiently large and finite k. The conclusion is that $V(x) = \sum_{i=1}^{s} \beta_i^*(x)$ is a Lyapunov function for $x(k) \in C_N$. Hence the closed loop system with the vertex control law is robustly asymptotically stable.   □

Example 3.4 Consider the discrete-time system in example 3.1,

$$x(k+1) = \begin{bmatrix} 1 & 1 \\ 0 & 1 \end{bmatrix} x(k) + \begin{bmatrix} 1 \\ 0.7 \end{bmatrix} u(k) \tag{3.50}$$

The constraints are,

$$-2 \le x_1 \le 2, \qquad -5 \le x_2 \le 5, \qquad -1 \le u \le 1 \tag{3.51}$$

Using Procedure 2.3, the set C_N is computed and depicted in Fig. 3.6.

The set of vertices of C_N is given by the matrix $V(C_N)$ below, together with the control matrix U_v,

$$\begin{cases} V(C_N) = \begin{bmatrix} 2.00 & 1.30 & -0.10 & -2.00 & -2.00 & -1.30 & 0.10 & 2.00 \\ 1.00 & 1.70 & 2.40 & 3.03 & -1.00 & -1.70 & -2.40 & -3.03 \end{bmatrix}, \\ U_v \quad = \begin{bmatrix} -1 & -1 & -1 & -1 & 1 & 1 & 1 & 1 \end{bmatrix} \end{cases} \tag{3.52}$$

The vertex control law over the state space partition is,

$$u(k) = \begin{cases} -0.25x_1(k) - 0.50x_2(k), & \text{if } x(k) \in C_N^{(1)} \text{ or } x(k) \in C_N^{(5)} \\ -0.33x_1(k) - 0.33x_2(k), & \text{if } x(k) \in C_N^{(2)} \text{ or } x(k) \in C_N^{(6)} \\ -0.21x_1(k) - 0.43x_2(k), & \text{if } x(k) \in C_N^{(3)} \text{ or } x(k) \in C_N^{(7)} \\ -0.14x_1(k) - 0.42x_2(k), & \text{if } x(k) \in C_N^{(4)} \text{ or } x(k) \in C_N^{(8)} \end{cases} \tag{3.53}$$

with

$$C_N^{(1)} = \left\{ x \in \mathbb{R}^2 : \begin{bmatrix} 1.00 & 0.00 \\ -0.45 & 0.89 \\ -0.83 & -0.55 \end{bmatrix} x \le \begin{bmatrix} 2.00 \\ 0.00 \\ 0.00 \end{bmatrix} \right\}$$

$$C_N^{(2)} = \left\{ x \in \mathbb{R}^2 : \begin{bmatrix} 0.71 & 0.71 \\ 0.45 & -0.89 \\ -0.79 & 0.61 \end{bmatrix} x \le \begin{bmatrix} 2.12 \\ 0.00 \\ 0.00 \end{bmatrix} \right\}$$

$$C_N^{(3)} = \left\{ x \in \mathbb{R}^2 : \begin{bmatrix} 0.79 & -0.61 \\ 0.45 & 0.89 \\ -1.00 & -0.04 \end{bmatrix} x \le \begin{bmatrix} 0.00 \\ 2.10 \\ 0.00 \end{bmatrix} \right\}$$

$$C_N^{(4)} = \left\{ x \in \mathbb{R}^2 : \begin{bmatrix} 1.00 & 0.04 \\ -0.83 & -0.55 \\ 0.32 & 0.95 \end{bmatrix} x \le \begin{bmatrix} 0.00 \\ 0.00 \\ 2.25 \end{bmatrix} \right\}$$

$$C_N^{(5)} = \left\{ x \in \mathbb{R}^2 : \begin{bmatrix} -1.00 & 0.00 \\ 0.45 & -0.89 \\ 0.83 & 0.55 \end{bmatrix} x \le \begin{bmatrix} 2.00 \\ 0.00 \\ 0.00 \end{bmatrix} \right\}$$

$$C_N^{(6)} = \left\{ x \in \mathbb{R}^2 : \begin{bmatrix} -0.71 & -0.71 \\ -0.45 & 0.89 \\ 0.79 & -0.61 \end{bmatrix} x \le \begin{bmatrix} 2.12 \\ 0.00 \\ 0.00 \end{bmatrix} \right\}$$

$$C_N^{(7)} = \left\{ x \in \mathbb{R}^2 : \begin{bmatrix} -0.79 & 0.61 \\ -0.45 & -0.89 \\ 1.00 & 0.04 \end{bmatrix} x \le \begin{bmatrix} 0.00 \\ 2.10 \\ 0.00 \end{bmatrix} \right\}$$

$$C_N^{(8)} = \left\{ x \in \mathbb{R}^2 : \begin{bmatrix} -1.00 & -0.04 \\ 0.83 & 0.55 \\ -0.32 & -0.95 \end{bmatrix} x \le \begin{bmatrix} 0.00 \\ 0.00 \\ 2.25 \end{bmatrix} \right\}$$

Figure 3.7 presents state trajectories of the closed loop system for different initial conditions.

For the initial condition $x(0) = [-2.0000 \ 3.0333]^T$, Fig. 3.8 shows the state, input trajectory, and the Lyapunov function $V(x) = \sum_{i=1}^s \beta_i^*$. As expected $V(x)$ is a positive and non-increasing function.

From Fig. 3.8(b), it is worth noticing that using the vertex controller, the control values are saturated only on the boundary of the set C_N, i.e. when $V(x) = 1$. And also the state trajectory at some moments is parallel to the boundary of the set C_N,

Fig. 3.7 State trajectories of the closed loop system for Example 3.4

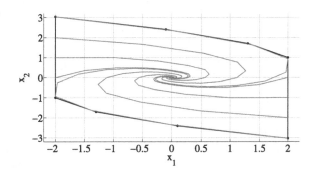

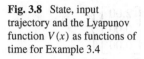

Fig. 3.8 State, input
trajectory and the Lyapunov
function $V(x)$ as functions of
time for Example 3.4

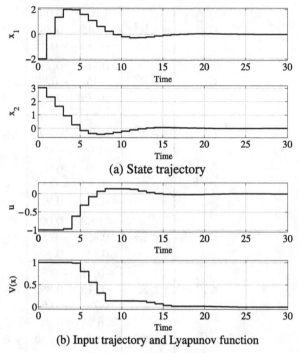

(a) State trajectory

(b) Input trajectory and Lyapunov function

i.e when $V(x)$ is constant. At these moments, the control values are also constant
due to the choice of the control values at the vertices of the set C_N.

Part II
Interpolating Control

Chapter 4
Interpolating Control—Nominal State Feedback Case

4.1 Problem Formulation

Consider the problem of regulating to the origin the following time-invariant linear discrete-time system,

$$x(k+1) = Ax(k) + Bu(k) \tag{4.1}$$

where $x(k) \in \mathbb{R}^n$ and $u(k) \in \mathbb{R}^m$ are respectively, the measurable state vector and the input vector. The matrices $A \in \mathbb{R}^{n \times n}$ and $B \in \mathbb{R}^{n \times m}$. Both $x(k)$ and $u(k)$ are subject to bounded polytopic constraints,

$$\begin{cases} x(k) \in X, \ X = \left\{ x \in \mathbb{R}^n : F_x x \le g_x \right\} \\ u(k) \in U, \ U = \left\{ u \in \mathbb{R}^m : F_u u \le g_u \right\} \end{cases} \quad \forall k \ge 0 \tag{4.2}$$

where the matrices F_x, F_u and the vectors g_x, g_u are assumed to be constant. The inequalities are taken element-wise. It is assumed that the pair (A, B) is stabilizable, i.e. all uncontrollable states have stable dynamics.

4.2 Interpolating Control via Linear Programming—Implicit Solution

Define a linear controller $K \in \mathbb{R}^{m \times n}$, such that,

$$u(k) = Kx(k) \tag{4.3}$$

asymptotically stabilizes the system (4.1) with some desired performance specifications. The details of such a synthesis procedure are not reproduced here, but we assume that feasibility is guaranteed. For the controller (4.3) using Procedure 2.1 or Procedure 2.2 the maximal invariant set $\Omega_{\max}$ can be computed as,

$$\Omega_{\max} = \left\{ x \in \mathbb{R}^n : F_o x \le g_o \right\} \tag{4.4}$$

H.-N. Nguyen, *Constrained Control of Uncertain, Time-Varying, Discrete-Time Systems*, 67
Lecture Notes in Control and Information Sciences 451,
DOI 10.1007/978-3-319-02827-9_4,
© Springer International Publishing Switzerland 2014

Fig. 4.1 Any state $x(k)$ can be decomposed as a convex combination of $x_v(k) \in C_N$ and $x_o(k) \in \Omega_{max}$

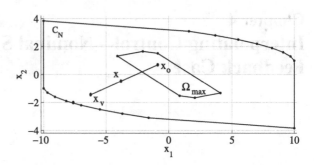

Furthermore with some given and fixed integer $N > 0$, based on Procedure 2.3 the controlled invariant set C_N can be found as,

$$C_N = \left\{ x \in \mathbb{R}^n : F_N x \leq g_N \right\} \tag{4.5}$$

such that all $x \in C_N$ can be steered into Ω_{max} in no more than N steps when a suitable control is applied. As in Sect. 3.4, the set C_N is decomposed as a sequence of simplices $C_N^{(j)}$, each formed by n vertices of C_N and the origin. For all $x(k) \in C_N^{(j)}$, the vertex controller

$$u(k) = K^{(j)} x(k), \tag{4.6}$$

with $K^{(j)}$ given in (3.38) asymptotically stabilizes the system (4.1), while the constraints (4.2) are fulfilled.

The main advantage of the vertex control scheme is the size of the domain of attraction, i.e. the set C_N. Clearly, C_N, that is the feasible domain for vertex control, might be as large as that of any other constrained control scheme. However, a weakness of vertex control is that the full control range is exploited only on the boundary of C_N in the state space, with progressively smaller control action when state approaches the origin. Hence the time to regulate the plant to the origin is often unnecessary long. A way to overcome this shortcoming is to switch to another, more aggressive, local controller, e.g. the controller (4.3), when the state reaches Ω_{max}. The disadvantage of this solution is that the control action becomes *nonsmooth* [94].

Here a method to overcome the nonsmooth control action [94] will be proposed. For this purpose, any state $x(k) \in C_N$ is decomposed as,

$$x(k) = c(k) x_v(k) + \left(1 - c(k)\right) x_o(k) \tag{4.7}$$

with $x_v \in C_N$, $x_o \in \Omega_{max}$ and $0 \leq c \leq 1$. Figure 4.1 illustrates such a decomposition.

Consider the following control law,

$$u(k) = c(k) u_v(k) + \left(1 - c(k)\right) u_o(k) \tag{4.8}$$

where $u_v(k)$ is the vertex control law (4.6) at $x_v(k)$ and $u_o(k) = K x_o(k)$ is the control law (4.3) in Ω_{max}.

Theorem 4.1 *For system* (4.1) *and constraints* (4.2), *the control law* (4.7), (4.8) *guarantees recursive feasibility for all initial states* $x(0) \in C_N$.

Proof For recursive feasibility, we have to prove that,

$$
\begin{cases}
F_u u(k) \leq g_u \\
x(k+1) = Ax(k) + Bu(k) \in C_N
\end{cases}
$$

for all $x(k) \in C_N$. For the input constraints,

$$
\begin{aligned}
F_u u(k) &= F_u \{c(k)u_v(k) + (1 - c(k))u_o(k)\} \\
&= c(k)F_u u_v(k) + (1 - c(k))F_u u_o(k) \\
&\leq c(k)g_u + (1 - c(k))g_u = g_u
\end{aligned}
$$

and for the state constraints,

$$
\begin{aligned}
x(k+1) &= Ax(k) + Bu(k) \\
&= A\{c(k)x_v(k) + (1 - c(k))x_o(k)\} + B\{c(k)u_v(k) + (1 - c(k))u_o(k)\} \\
&= c(k)\{Ax_v(k) + Bu_v(k)\} + (1 - c(k))\{Ax_o(k) + Bu_o(k)\}
\end{aligned}
$$

Since $Ax_v(k) + Bu_v(k) \in C_N$ and $Ax_o(k) + Bu_o(k) \in \Omega_{\max} \subseteq C_N$, it follows that $x(k+1) \in C_N$. ☐

Since the controller (4.3) is designed to give specified unconstrained performance in $\Omega_{\max}$, it might be desirable to have $u(k)$ in (4.8) as close as possible to it also outside $\Omega_{\max}$. This can be achieved by minimizing c,

$$
c^* = \min_{x_v, x_o, c} \{c\} \tag{4.9}
$$

subject to

$$
\begin{cases}
F_N x_v \leq g_N, \\
F_o x_o \leq g_o, \\
cx_v + (1 - c)x_o = x, \\
0 \leq c \leq 1
\end{cases}
$$

Denote $r_v = cx_v \in \mathbb{R}^n$, $r_o = (1 - c)x_o \in \mathbb{R}^n$. Since $x_v \in C_N$ and $x_o \in \Omega_{\max}$, it follows that $r_v \in cC_N$ and $r_o \in (1 - c)\Omega_{\max}$ or equivalently

$$
\begin{cases}
F_N r_v \leq cg_N \\
F_o r_o \leq (1 - c)g_o
\end{cases}
$$

Hence the nonlinear optimization problem (4.9) is transformed into the following linear programming problem,

$$c^* = \min_{r_v, c}\{c\} \qquad (4.10)$$

subject to

$$\begin{cases} F_N r_v \leq c g_N, \\ F_o(x - r_v) \leq (1 - c)g_o, \\ 0 \leq c \leq 1 \end{cases}$$

Remark 4.1 If one would like to maximize c, it is obvious that $c = 1$ for all $x \in C_N$. In this case the controller (4.7), (4.8) becomes the vertex controller.

Theorem 4.2 *The control law* (4.7), (4.8), (4.10) *guarantees asymptotic stability for all initial states* $x(0) \in C_N$.

Proof First of all we will prove that all solutions starting in $C_N \setminus \Omega_{max}$ will reach Ω_{max} in *finite time*. For this purpose, consider the following non-negative function,

$$V(x) = c^*(x), \quad \forall x \in C_N \setminus \Omega_{max} \qquad (4.11)$$

$V(x)$ is a candidate Lyapunov function. After solving the LP problem (4.10) and applying (4.7), (4.8), one obtains, for $x(k) \in C_N \setminus \Omega_{max}$,

$$\begin{cases} x(k) = c^*(k)x_v^*(k) + (1 - c^*(k))x_o^*(k) \\ u(k) = c^*(k)u_v(k) + (1 - c^*(k))u_o(k) \end{cases}$$

It follows that,

$$\begin{aligned} x(k + 1) &= Ax(k) + Bu(k) \\ &= c^*(k)x_v(k + 1) + (1 - c^*(k))x_o(k + 1) \end{aligned}$$

where

$$\begin{cases} x_v(k + 1) = Ax_v^*(k) + Bu_v(k) \in C_N \\ x_o(k + 1) = Ax_o^*(k) + Bu_o(k) \in \Omega_{max} \end{cases}$$

Hence $c^*(k)$ is a feasible solution for the LP problem (4.10) at time $k + 1$. By solving (4.10) at time $k + 1$, one gets the optimal solution, namely

$$x(k + 1) = c^*(k + 1)x_v^*(k + 1) + (1 - c^*(k + 1))x_o^*(k + 1)$$

where $x_v^*(k + 1) \in C_N$ and $x_o^*(k + 1) \in \Omega_{max}$. It follows that $c^*(k + 1) \leq c^*(k)$ and $V(x)$ is non-increasing.

Using the vertex controller, an interpolation between a point of C_N and the origin is obtained. Conversely using the controller (4.7), (4.8), (4.10) an interpolation is constructed between a point of C_N and a point of Ω_{max} which in turn contains the

Algorithm 4.1 Interpolating control—Implicit solution

1. Measure the current state $x(k)$.
2. Solve the LP problem (4.10).
3. Compute u_{rv} in (4.12) by determining to which simplex r_v^* belongs and using (3.38).
4. Implement as input the control signal (4.12).
5. Wait for the next time instant $k := k + 1$.
6. Go to step 1 and repeat.

origin as an interior point. This last property proves that the vertex controller is a feasible choice for the interpolation scheme (4.7), (4.8), (4.10). Hence it follows that,

$$c^*(k) \leq \sum_{i=1}^{s} \beta_i^*(k)$$

for any $x(k) \in C_N$, with $\beta_i^*(k)$ obtained in (3.46), Sect. 3.4.

Since the vertex controller is asymptotically stabilizing, the state reaches any bounded set around the origin in finite time. In our case this property will imply that using the controller (4.7), (4.8), (4.10) the state of the closed loop system reaches Ω_{max} in *finite time* or equivalently that there exists a finite k such that $c^*(k) = 0$.

The proof is complete by noting that inside Ω_{max}, the LP problem (4.10) has the trivial solution $c^* = 0$. Hence the controller (4.7), (4.8), (4.10) becomes the local controller (4.3). The feasible stabilizing controller $u(k) = Kx(k)$ is *contractive*, and thus the interpolating controller assures asymptotic stability for all $x \in C_N$.   □

The control law (4.7), (4.8), (4.10) obtained by solving on-line the LP problem (4.10) is called Implicit Interpolating Control.

Since $r_v^*(k) = c^*(k)x_v^*(k)$ and $r_o^*(k) = (1 - c^*(k))x_o^*(k)$, it follows that,

$$u(k) = u_{rv}(k) + u_{ro}(k) \tag{4.12}$$

where $u_{rv}(k)$ is the vertex control law at $r_v^*(k)$ and $u_{ro}(k) = Kr_o^*(k)$.

Remark 4.2 Note that at each time instant Algorithm 4.1 requires the solutions of two LP problems, one is (4.10) of dimension $n + 1$, the other is to determine to which simplex r_v^* belongs.

Example 4.1 Consider the following time-invariant linear discrete-time system,

$$x(k + 1) = \begin{bmatrix} 1 & 1 \\ 0 & 1 \end{bmatrix} x(k) + \begin{bmatrix} 1 \\ 0.3 \end{bmatrix} u(k) \tag{4.13}$$

The constraints are,

$$-10 \leq x_1(k) \leq 10, \qquad -5 \leq x_2(k) \leq 5, \qquad -1 \leq u(k) \leq 1 \tag{4.14}$$

The local controller is chosen as a linear quadratic (LQ) controller with weighting matrices $Q = I$ and $R = 1$, giving the state feedback gain,

$$K = [-0.5609 \quad -0.9758] \tag{4.15}$$

The sets Ω_{max} and C_N with $N = 14$ are shown in Fig. 4.1. Note that $C_{14} = C_{15}$ is the maximal controlled invariant set. Ω_{max} is presented in minimal normalized half-space representation as,

$$\Omega_{max} = \left\{ x \in \mathbb{R}^2 : \begin{bmatrix} 0.1627 & -0.9867 \\ -0.1627 & 0.9867 \\ -0.1159 & -0.9933 \\ 0.1159 & 0.9933 \\ -0.4983 & -0.8670 \\ 0.4983 & 0.8670 \end{bmatrix} x \leq \begin{bmatrix} 1.9746 \\ 1.9746 \\ 1.4115 \\ 1.4115 \\ 0.8884 \\ 0.8884 \end{bmatrix} \right\} \tag{4.16}$$

The set of vertices of C_N is given by the matrix $V(C_N)$, together with the corresponding control matrix U_v,

$$V(C_N) = [V_1 \quad -V_1], \qquad U_v = [U_1 \quad -U_1] \tag{4.17}$$

where

$$V_1 = \begin{bmatrix} 10.0000 & 9.7000 & 9.1000 & 8.2000 & 7.0000 & 5.5000 & 3.7000 & 1.6027 & -10.0000 \\ 1.0000 & 1.3000 & 1.6000 & 1.9000 & 2.2000 & 2.5000 & 2.8000 & 3.0996 & 3.8368 \end{bmatrix},$$

$$U_1 = \begin{bmatrix} -1 & -1 & -1 & -1 & -1 & -1 & -1 & -1 & 1 \end{bmatrix}$$

The state space partition of vertex control is shown in Fig. 4.2(a). Using the implicit interpolating controller, Fig. 4.2(b) presents state trajectories of the closed loop system for different initial conditions.

For the initial condition $x(0) = [-2.0000 \ 3.3284]^T$, Fig. 4.3 shows the state and input trajectories for the implicit interpolating controller (solid). As a comparison, we take MPC, based on quadratic programming, where an LQ criterion is optimized, with identity weighting matrices. Hence the set Ω_{max} for the local unconstrained control is identical for the MPC solution and for the implicit interpolating controller. The prediction horizon for the MPC was chosen to be 14 to match the controlled invariant set C_{14} used for the implicit interpolating controller. Figure 4.3 shows the state and input trajectories obtained for the implicit MPC (dashed).

Using the tic/toc function of Matlab 2011b, the computational burdens of interpolating control and MPC were compared. The result is shown in Table 4.1

Table 4.1 Durations [ms] of the on-line computations during one sampling interval for interpolating control and MPC, respectively for Example 4.1

	Computational time
Implicit interpolating control	0.7652
Implicit QP-MPC	4.6743

Fig. 4.2 State space partition of vertex control and state trajectories for Example 4.1

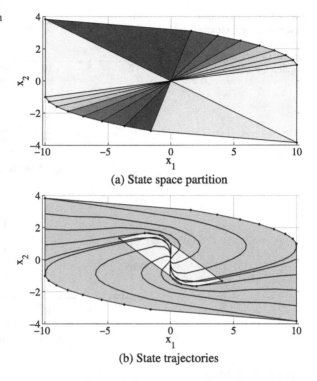

(a) State space partition

(b) State trajectories

As a final analysis element, Fig. 4.4 presents the interpolating coefficient $c^*(k)$. It is interesting to note that $c^*(k) = 0$, $\forall k \geq 15$ indicating that from time instant $k = 15$, the state of the closed loop system is in $\Omega_{\max}$, and consequently is *optimal* in the MPC cost function terms. The monotonic decrease and the positivity confirms the Lyapunov interpretation given in the present section.

4.3 Interpolating Control via Linear Programming—Explicit Solution

The structural implication of the LP problem (4.10) is investigated in this section.

4.3.1 Geometrical Interpretation

Let $\partial(\cdot)$ denotes the boundary of the corresponding set $(\cdot)$. The following theorem holds

Theorem 4.3 *For all $x \in C_N \setminus \Omega_{\max}$, the solution of the LP problem* (4.10) *satisfies* $x_v^* \in \partial C_N$ *and* $x_o^* \in \partial \Omega_{\max}$.

Fig. 4.3 State and input
trajectories for Example 4.1
for implicit interpolating
control (*solid*), and for
implicit QP-MPC (*dashed*)

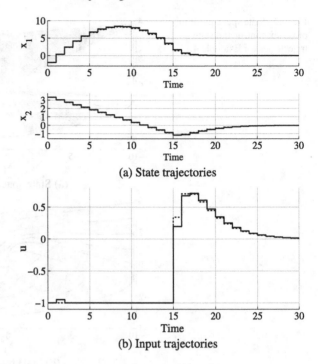

(a) State trajectories

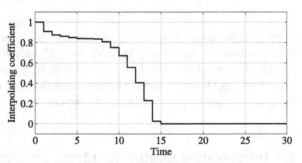

(b) Input trajectories

Fig. 4.4 Interpolating
coefficient c^* as a function of
time example 4.1

Proof Consider $x \in C_N \setminus \Omega_{\max}$, with a particular convex combination

$$x = cx_v + (1-c)x_o$$

where $x_v \in C_N$ and $x_o \in \Omega_{\max}$. If x_o is strictly inside $\Omega_{\max}$, one can set $\tilde{x}_o = \partial\Omega_{\max} \cap \overline{x, x_o}$, i.e. $\tilde{x}_o$ is the intersection between $\partial\Omega_{\max}$ and the line segment connecting x and x_o, see Fig. 4.5. Apparently, x can be expressed as the convex combination of x_v and $\tilde{x}_o$, i.e.

$$x = \tilde{c}x_v + (1-\tilde{c})\tilde{x}_o$$

with $\tilde{c} < c$, since x is closer to $\tilde{x}_o$ than to x_o. So (4.10) leads to $\{c^*, x_v^*, x_o^*\}$ with $x_o^* \in \partial\Omega_{\max}$.

Fig. 4.5 Graphical
illustration for the proof of
Theorem 4.3

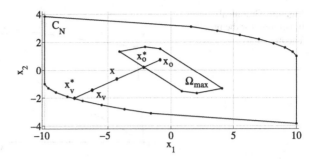

One the other hand, if x_v is strictly inside C_N, one can set $\tilde{x}_v = \partial C_N \cap \overrightarrow{x, x_v}$,
i.e. $\tilde{x}_v$ is the intersection between ∂C_N and the ray starting from x through x_v, see
Fig. 4.5. Again, x can be written as the convex combination of $\tilde{x}_v$ and x_o, i.e.

$$x = \tilde{c}\tilde{x}_v + (1 - \tilde{c})x_o$$

with $\tilde{c} < c$, since x is further from $\tilde{x}_v$ than from x_v. This leads to the conclusion that
for the optimal solution $\{c^*, x_v^*, x_o^*\}$ we have $x_v^* \in \partial P_N$.   □

Theorem 4.3 states that for all $x \in C_N \setminus \Omega_{max}$, the interpolating coefficient c
is minimal if and only if x is written as a convex combination of two points, one
belonging to C_N and the other to $\partial \Omega_{max}$. It is obvious that for $x \in \Omega_{max}$, the LP
problem (4.10) has the trivial solution $c^* = 0$ and thus $x_v^* = 0$ and $x_o^* = x$.

Theorem 4.4   *For all $x \in C_N \setminus \Omega_{max}$, the convex combination $x = cx_v + (1 - c)x_o$
gives the smallest value of c if the ratio $\frac{\|x_v - x\|}{\|x - x_o\|}$ is maximal, where $\|\cdot\|$ denotes the
Euclidean vector norm.*

Proof   It holds that

$$x = cx_v + (1 - c)x_o$$
$$\Rightarrow \quad x_v - x = x_v - cx_v - (1 - c)x_o = (1 - c)(x_v - x_o)$$

consequently

$$\|x_v - x\| = (1 - c)\|x_v - x_o\| \tag{4.18}$$

Analogously, one obtains

$$\|x - x_o\| = c\|x_v - x_o\| \tag{4.19}$$

Combining (4.18) and (4.19) and the fact that $c \neq 0$ for all $x \in C_N \setminus \Omega_{max}$, one gets

$$\frac{\|x_v - x\|}{\|x - x_o\|} = \frac{(1 - c)\|x_v - x_o\|}{c\|x_v - x_o\|} = \frac{1}{c} - 1$$

$c > 0$ is minimal if and only if $\frac{1}{c} - 1$ is maximal, or equivalently $\frac{\|x_v - x\|}{\|x - x_o\|}$ is maximal.   □

Fig. 4.6 Graphical
illustration for the proof of
Theorem 4.5

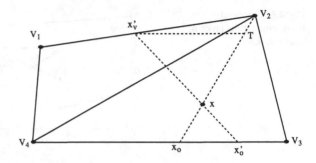

4.3.2 Analysis in $\mathbb{R}^2$

In this subsection an analysis of the optimization problem (4.9) in the $\mathbb{R}^2$ parameter space is presented with reference to Fig. 4.6. The discussion is insightful in what concerns the properties of the partition in the explicit solution. The problem considered here is to decompose the polyhedral X_{1234} such that the explicit solution $c^* = \min\{c\}$ is given in the decomposed cells.

For illustration we will consider four points V_1, V_2, V_3, V_4, and any point $x \in \text{Conv}(V_1, V_2, V_3, V_4)$. This schematic view can be generalized to any pair of faces of C_N and Ω_{max}. Denote V_{ij} as the interval connecting V_i and V_j for $i, j = 1, \ldots, 4$. The problem is reduced to the expression of a convex combination $x = cx_v + (1 - c)x_o$, where $x_v \in V_{12} \subset \partial C_N$ and $x_o \in V_{34} \subset \partial \Omega_{max}$ providing the minimal value of c.

Without loss of generality, suppose that the distance from V_2 to V_{34} is greater than the distance from V_1 to V_{34}, or equivalently the distance from V_4 to V_{12} is smaller than the distance from V_3 to V_{12}.

Theorem 4.5 *Under the condition that the distance from V_2 to V_{34} is greater than the distance from V_1 to V_{34}, or equivalently the distance from V_4 to V_{12} is smaller than the distance from V_3 to V_{12}, the decomposition of the polytope V_{1234}, $V_{1234} = V_{124} \cup V_{234}$ is the result of the minimization of the interpolating coefficient c.*

Proof Without loss of generality, suppose that $x \in V_{234}$. x can be decomposed as,

$$x = cV_2 + (1 - c)x_o \tag{4.20}$$

where $x_o \in V_{34}$, see Fig. 4.6. Another possible decomposition is

$$x = c'x_v' + (1 - c')x_o' \tag{4.21}$$

where x_v' belongs to V_{34} and x_o' belongs to V_{12}.

Clearly, if the distance from V_2 to V_{34} is greater than the distance from V_1 to V_{34} then the distance from V_2 to V_{34} is greater than the distance from any point in V_{12} to V_{34}. Consequently, there exists the point T in the ray, starting from V_2 through

x such that the distance from T to V_{34} is equal to the distance from x'_v to V_{34}. It follows that the line connecting T and x'_v is parallel to X_{34}, see Fig. 4.6.

Using Basic Proportionality Theorem, one has

$$\frac{\|x - x'_v\|}{\|x - x'_o\|} = \frac{\|x - T\|}{\|x - x_o\|} \tag{4.22}$$

by using Theorem 4.4 and since

$$\frac{\|x - T\|}{\|x - x_o\|} < \frac{\|x - V_2\|}{\|x - x_o\|}$$

it follows that $c < c'$. □

Theorem 4.5 states that the minimal value of the interpolating coefficient c is found with the help of the decomposition of V_{1234} as $V_{1234} = V_{124} \cup V_{234}$.

Remark 4.3 Clearly, if V_{12} is parallel to V_{34}, then any convex combination $x = cx_v + (1 - c)x_o$ gives the same value of c. Hence the partition may not be unique.

Remark 4.4 As a consequence of Theorem 4.5, it is clear that the region $C_N \setminus \Omega_{max}$ can be subdivided into partitions (cells) as follows,

- For each facet of the set Ω_{max}, one has to find the furthest point on ∂C_N on the same side of the origin as the facet of Ω_{max}. A polyhedral cell is obtained as the convex hull of that facet of Ω_{max} and the furthest point in C_N. By the bounded polyhedral structure of C_N, the existence of some vertex of C_N as the furthest point is guaranteed.
- On the other hand, for each facet of C_N, one has to find the closest point on $\partial \Omega_{max}$ on the same side of the origin as the facet of C_N. A polyhedral cell is obtained as the convex hull of that facet of C_N and the closest point in Ω_{max}. Again by the bounded polyhedral structure of Ω_{max}, the existence of some vertex Ω_{max} as the closest point is guaranteed.

Remark 4.5 Clearly, in $\mathbb{R}^2$, the state space partition according to Remark 4.4 cover the entire set C_N, see e.g. Fig. 4.7. However in $\mathbb{R}^n$, that is not necessarily the case as shown in the following example. Let C_N and Ω_{max} be given by the vertex representations, displayed in Fig. 4.8(a),

$$C_N = \text{Conv} \left\{ \begin{bmatrix} -4 \\ 0 \\ 0 \end{bmatrix}, \begin{bmatrix} 4 \\ 4 \\ 4 \end{bmatrix}, \begin{bmatrix} 4 \\ -4 \\ 0 \end{bmatrix}, \begin{bmatrix} 4 \\ 4 \\ -4 \end{bmatrix} \right\}$$

$$\Omega_{max} = \text{Conv} \left\{ \begin{bmatrix} 1 \\ 0 \\ 0 \end{bmatrix}, \begin{bmatrix} -0.5 \\ -0.5 \\ -0.5 \end{bmatrix}, \begin{bmatrix} -0.5 \\ 0.5 \\ 0 \end{bmatrix}, \begin{bmatrix} -0.5 \\ -0.5 \\ 0.5 \end{bmatrix} \right\}$$

Fig. 4.7 Simplex based
decomposition as an explicit
solution of the LP problem
(4.10)

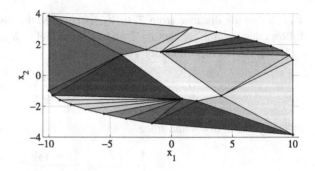

Fig. 4.8 Graphical
illustration for Remark 4.5.
The partition is obtained by
two vertices of the inner set
Ω_{max} and two vertices of the
outer set C_N

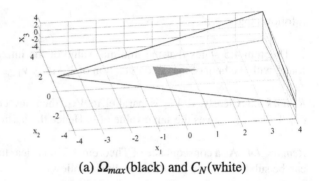

(a) Ω_{max}(black) and C_N(white)

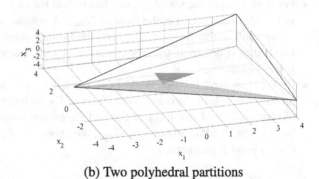

(b) Two polyhedral partitions

By solving the parametric linear programming problem (4.10) with respect to x, the
state space partition is obtained [19]. Figure 4.8(b) shows two polyhedral partitions
of the state space partition. The black set is Ω_{max}. The gray set is the convex hull of
two vertices of Ω_{max} and *two vertices* of C_N.

In conclusion, in $\mathbb{R}^n$ for all $x \in C_N \setminus \Omega_{max}$, the smallest value c will be reached
when $C_N \setminus \Omega_{max}$ is decomposed into polytopes with vertices both on ∂C_N and
$\partial\Omega_{max}$. These polytopes can be further decomposed into simplices, each formed by
r vertices of C_N and $n - r + 1$ vertices of Ω_{max} where $1 \leq r \leq n$.

4.3.3 Explicit Solution

Theorem 4.6 *For all $x \in C_N \setminus \Omega_{\max}$, the controller* (4.7), (4.8), (4.10) *is a piecewise affine state feedback law defined over a partition of $C_N \setminus \Omega_{\max}$ into simplices. The controller gains are obtained by linear interpolation of the control values at the vertices of simplices.*

Proof Suppose that x belongs to a simplex formed by n vertices $\{v_1, v_2, \ldots, v_n\}$ of C_N and the vertex v_o of $\Omega_{\max}$. The other cases of $n+1$ vertices distributed in a different manner between C_N and $\Omega_{\max}$ can be treated similarly.

In this case, x can be expressed as,

$$x = \sum_{i=1}^{n} \beta_i v_i + \beta_{n+1} v_o \tag{4.23}$$

where

$$\sum_{i=1}^{n+1} \beta_i = 1, \quad \beta_i \geq 0 \tag{4.24}$$

Given that $n+1$ linearly independent vectors define a non-empty simplex, let the invertible $(n+1) \times (n+1)$ matrix be

$$T_s = \begin{bmatrix} v_1 & v_2 & \cdots & v_n & v_o \\ 1 & 1 & \cdots & 1 & 1 \end{bmatrix} \tag{4.25}$$

Using (4.23), (4.24), (4.25), the interpolating coefficients β_i with $i = 1, 2, \ldots, n+1$ are defined uniquely as,

$$\begin{bmatrix} \beta_1 & \beta_2 & \cdots & \beta_n & \beta_{n+1} \end{bmatrix}^T = T_s^{-1} \begin{bmatrix} x \\ 1 \end{bmatrix} \tag{4.26}$$

On the other hand, from (4.7),

$$x = cx_v + (1-c)x_o,$$

Due to the uniqueness of (4.23), $\beta_{n+1} = 1 - c$ and

$$x_v = \sum_{i=1}^{n} \frac{\beta_i}{c} v_i$$

The Vertex Controller (3.46) gives

$$u_v = \sum_{i=1}^{n} \frac{\beta_i}{c} u_i$$

where u_i are an admissible control value at v_i, $i = 1, 2, \ldots, n$. Therefore

$$u = cu_v + (1 - c)u_o = \sum_{i=1}^{n} \beta_i u_i + \beta_{n+1} u_o.$$

with $u_o = K x_o$. Together with (4.26), one obtains

$$u = \begin{bmatrix} u_1 & u_2 & \ldots & u_n & u_o \end{bmatrix} \begin{bmatrix} \beta_1 & \beta_2 & \ldots & \beta_n & \beta_{n+1} \end{bmatrix}^T$$

$$= \begin{bmatrix} u_1 & u_2 & \ldots & u_n & u_o \end{bmatrix} T_s^{-1} \begin{bmatrix} x \\ 1 \end{bmatrix}$$

$$= Lx + v$$

where the matrix $L \in \mathbb{R}^{m \times n}$ and the vector $v \in \mathbb{R}^m$ are defined by,

$$\begin{bmatrix} L & v \end{bmatrix} = \begin{bmatrix} u_1 & u_2 & \ldots & u_n & u_o \end{bmatrix} T_s^{-1}$$

Hence for all $x \in C_N \setminus \Omega_{\max}$ the controller (4.7), (4.8), (4.10) is a piecewise affine state feedback law. □

It is interesting to note that the interpolation between the *piecewise linear* Vertex Controller and the *linear* controller in $\Omega_{\max}$ give rise to a *piecewise affine* controller. This is not completely unexpected since (4.10) is a multi-parametric linear program with respect to x.

As in MPC, the number of cells can be reduced by merging those with identical control laws [45].

Remark 4.6 It can be observed that Algorithm 4.2 uses only the information about the state space partition of the explicit solution of the LP problem (4.10). The explicit form of c^*, r_v^* and r_o^* as a piecewise affine function of the state is not used.

Clearly, the simplex-based partition over $C_N \setminus \Omega_{\max}$ in step 2 might be very complex. Also the fact, that for all facets of $\Omega_{\max}$ the local controller is of the form $u = Kx$, is not exploited. In addition, as practice usually shows, for each facet of C_N, the vertex controller is usually constant. In these cases, the complexity of the explicit interpolating controller (4.7), (4.8), (4.10) might be reduced as follows.

Consider the case when the state space partition CR of $C_N \setminus \Omega_{\max}$ is formed by one vertex x_v of C_N and one facet F_o of $\Omega_{\max}$. Note that from Remark 4.4 such a partition always exists as an explicit solution to the LP problem (4.10). For all $x \in CR$ it follows that

$$x = c^* x_v^* + (1 - c^*) x_o^* = c^* x_v^* + r_o^*$$

with $x_o^* \in F_o$ and $r_o^* = (1 - c^*) x_o^*$.

Algorithm 4.2 Interpolating control—Explicit solution

Input: The sets C_N, Ω_{max}, the optimal feedback controller $u = Kx$ in Ω_{max} and the control values at the vertices of C_N.
Output: The piecewise affine control law over the partitions of C_N.

1. Solve the LP (4.10) by using explicit multi-parametric linear programming. As a result, one obtains the state space partition of C_N.
2. Decompose each polyhedral partition of $C_N \setminus \Omega_{max}$ in a sequence of simplices, each formed by r vertices of C_N and $n - z + 1$ vertex of Ω_{max}, where $1 \leq z \leq n$. The result is a the state space partition over $C_N \setminus \Omega_{max}$ in the form of simplices CR_i.
3. In each simplex $CR_i \subset C_N \setminus \Omega_{max}$ the control law is defined as,

$$u(x) = L_i x + v_i \tag{4.27}$$

where $L_i \in \mathbb{R}^{m \times n}$ and $v_i \in \mathbb{R}^m$ are defined as

$$\begin{bmatrix} L_i & v_i \end{bmatrix} = \begin{bmatrix} u_1^{(i)} & u_2^{(i)} & \cdots & u_{n+1}^{(i)} \end{bmatrix} \begin{bmatrix} v_1^{(i)} & v_2^{(i)} & \cdots & v_{n+1}^{(i)} \\ 1 & 1 & \cdots & 1 \end{bmatrix}^{-1} \tag{4.28}$$

with $\{v_1^{(i)}, v_2^{(i)}, \ldots v_{n+1}^{(i)}\}$ are vertices of CR_i that defines a full-dimensional simplex and $\{u_1^{(i)}, u_2^{(i)}, \ldots u_{n+1}^{(i)}\}$ are the corresponding control values at the vertices.

Let $u_v \in \mathbb{R}^m$ be an admissible control value at x_v and denote the explicit solution of c^* and r_o^* to the LP problem (4.10) for all $x \in CR$ as,

$$\begin{cases} c^* = L_c x + v_c \\ r_o^* = L_o x + v_o \end{cases} \tag{4.29}$$

where L_c, v_c and L_o, v_o are matrices of appropriate dimensions. The control value for $x \in CR$ is computed as,

$$u = c^* u_v + (1 - c^*) K x_o^* = c^* u_v + K r_o^* \tag{4.30}$$

By substituting (4.29) into (4.30), one obtains

$$u = u_v (L_c x + v_c) + K (L_o x + v_o)$$

or, equivalently

$$u = (u_v L_c + K L_o) x + (u_v v_c + K v_o) \tag{4.31}$$

The fact that the control value is a piecewise affine function of state is confirmed. Clearly, the complexity of the explicit solution with the control law (4.31) is lower than the complexity of the explicit solution with the simplex based partition, since

Fig. 4.9 Graphical
illustration for the proof of
Theorem 4.7

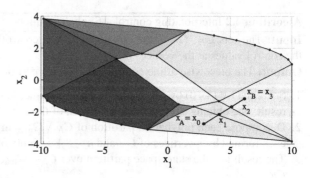

one does not have to divide up the facets of $\Omega_{\max}$ (and facets of C_N, in the case
when the vertex control for such facets is constant) into a set of simplices.

4.3.4 Qualitative Analysis

Theorem 4.7 below shows the Lipschitz continuity of the control law based on linear
programming (4.7), (4.8), (4.10).

Theorem 4.7 *The explicit interpolating control law* (4.7), (4.8), (4.10) *obtained by
using Algorithm* 4.2 *is continuous and Lipschitz continuous with Lipschitz constant*
$M = \max_i \|L_i\|$, *where i ranges over the set of indices of partitions and* $\|L_i\|$ *is
defined in* (4.28).

Proof The explicit interpolating controller might be discontinuous only on the
boundary of polyhedral cells CR_i. Suppose that x belongs to the intersection of
s cells CR_j, $j = 1, 2, \ldots, s$.

For CR_j, as in (4.23), the state x can be expressed as,

$$x = \beta_1^{(j)} v_1^{(j)} + \beta_2^{(j)} v_2^{(j)} + \cdots + \beta_{n+1}^{(j)} v_{n+1}^{(j)}$$

where $\sum_{i=1}^{n+1} \beta_i^{(j)} = 1$, $0 \le \beta_i^{(j)} \le 1$ and $v_i^{(j)}$, $i = 1, 2, \ldots, n + 1$ are the vertices
of CR_j, $j = 1, 2, \ldots, s$. It is clear that the only nonzero entries of the interpolating
coefficients $\{\beta_1^{(j)}, \ldots, \beta_{n+1}^{(j)}\}$ are those corresponding to the vertices that belong to
the intersection. Therefore

$$u = \beta_1^{(j)} u_1^{(j)} + \cdots + \beta_{n+1}^{(j)} u_{n+1}^{(j)}$$

is equal for all $j = 1, 2, \ldots, s$.

For the Lipschitz continuity property, for any two points x_A and x_B in C_N, there
exist $r + 1$ points $x_0, x_1, \ldots, x_r$ that lie on the line segment, connecting x_A and x_B,
and such that $x_A = x_0$, $x_B = x_r$ and $(x_{i-1}, x_i) = \overline{x_A, x_B} \cap \partial CR_i$, i.e. (x_{i-1}, x_i) is the
intersection between the line connecting x_A, x_B and the boundary of some critical

Fig. 4.10 Lyapunov function and Lyapunov level curves for the interpolating controller for Example 4.2

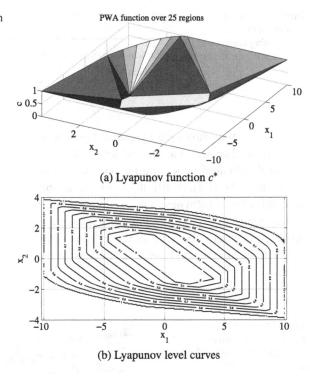

(a) Lyapunov function c^*

(b) Lyapunov level curves

region CR_i, see Fig. 4.9. Due to the continuity property, proved above, of the control law (4.27), one has,

$$\left\|(L_A x_A + v_A) - (L_B x_B + v_B)\right\|$$

$$= \left\|(L_0 x_0 + v_0) - (L_0 x_1 + v_0) + (L_1 x_1 + v_1) - \cdots - (L_r x_r + v_r)\right\|$$

$$= \left\|L_0 x_0 - L_0 x_1 + L_1 x_1 - \cdots - L_r x_r\right\|$$

$$\leq \sum_{i=1}^{r} \left\|L_{i-1}(x_i - x_{i-1})\right\| \leq \sum_{k=1}^{r} \left\|L_{i-1}\right\| \left\|(x_i - x_{i-1})\right\|$$

$$\leq \max_{k} \left\{\left\|L_{i-1}\right\|\right\} \sum_{i=1}^{r} \left\|(x_i - x_{i-1})\right\| = M \|x_A - x_B\|$$

where the last equality holds, since the points x_i with $k = 0, 1, \ldots, r$ are aligned. $\square$

Example 4.2 We consider now the explicit interpolating controller for Example 4.1. Using Algorithm 4.2, the state space partition is obtained in Fig. 4.7. Merging the regions with identical control laws, the reduced state space partition is obtained in Fig. 4.9.

Table 4.2 Number of regions for explicit interpolating control and for explicit MPC for Example 4.2

	Before merging	After merging
Explicit interpolating control	25	11
Explicit MPC	127	97

Figure 4.10(a) shows the Lyapunov function as a piecewise affine function of state. It is well known[1] that the level sets of the Lyapunov function for vertex control are simply obtained by scaling the boundary of the set C_N. For the interpolating controller (4.7), (4.8), (4.10), the level sets of the Lyapunov function $V(x) = c^*$ depicted in Fig. 4.10(b) have a more complicated form and generally are not parallel to the boundary of C_N. From Fig. 4.10, it can be observed that the Lyapunov level sets $V(x) = c^*$ have the outer set C_N as an external level set (for $c^* = 1$). The inner level sets change the polytopic shape in order to approach the boundary of the inner set Ω_{max}.

Fig. 4.11 State space partition before and after merging for Example 4.2 using explicit MPC

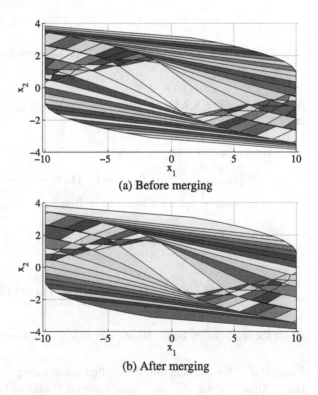

(a) Before merging

(b) After merging

---

[1] See Sect. 3.4.

The control law over the state space partition is,

$$
u(k) = \begin{cases}
-1 & \text{if} \begin{bmatrix} 0.45 & 0.89 \\ 0.24 & 0.97 \\ 0.16 & 0.99 \\ -0.55 & 0.84 \\ 0.14 & 0.99 \\ -0.50 & -0.87 \\ 0.20 & 0.98 \\ 0.32 & 0.95 \\ 0.37 & -0.93 \\ 0.70 & 0.71 \end{bmatrix} x(k) \leq \begin{bmatrix} 5.50 \\ 3.83 \\ 3.37 \\ 1.75 \\ 3.30 \\ -0.89 \\ 3.53 \\ 4.40 \\ 2.73 \\ 7.78 \end{bmatrix} \\[2mm]
-0.38x_1(k) + 0.59x_2(k) - 2.23 & \text{if} \begin{bmatrix} 0.54 & -0.84 \\ -0.37 & 0.93 \\ -0.12 & -0.99 \end{bmatrix} x(k) \leq \begin{bmatrix} -1.75 \\ 2.30 \\ -1.41 \end{bmatrix} \\[2mm]
-0.02x_1(k) - 0.32x_2(k) + 0.02 & \text{if} \begin{bmatrix} 0.37 & -0.93 \\ 0.06 & 1.00 \\ -0.26 & -0.96 \end{bmatrix} x(k) \leq \begin{bmatrix} -2.30 \\ 3.20 \\ -1.06 \end{bmatrix} \\[2mm]
-0.43x_1(k) - 1.80x_2(k) + 1.65 & \text{if} \begin{bmatrix} 0.16 & -0.99 \\ 0.26 & 0.96 \\ -0.39 & -0.92 \end{bmatrix} x(k) \leq \begin{bmatrix} -1.97 \\ 1.06 \\ 0.38 \end{bmatrix} \\[2mm]
0.16x_1(k) - 0.41x_2(k) + 2.21 & \text{if} \begin{bmatrix} 0.39 & 0.92 \\ -1.00 & 0 \\ 0.37 & -0.93 \end{bmatrix} x(k) \leq \begin{bmatrix} -0.38 \\ 10.00 \\ -2.73 \end{bmatrix} \\[2mm]
1 & \text{if} \begin{bmatrix} -0.14 & -0.99 \\ -0.37 & 0.93 \\ -0.24 & -0.97 \\ -0.71 & -0.71 \\ -0.45 & -0.89 \\ -0.32 & -0.95 \\ -0.20 & -0.98 \\ -0.16 & -0.99 \\ 0.50 & 0.87 \\ 0.54 & -0.84 \end{bmatrix} x(k) \leq \begin{bmatrix} 3.30 \\ 2.73 \\ 3.83 \\ 7.78 \\ 5.50 \\ 4.40 \\ 3.53 \\ 3.37 \\ -0.89 \\ 1.75 \end{bmatrix} \\[2mm]
-0.38x_1(k) + 0.59x_2(k) + 2.23 & \text{if} \begin{bmatrix} 0.12 & 0.99 \\ 0.37 & -0.93 \\ -0.54 & 0.84 \end{bmatrix} x(k) \leq \begin{bmatrix} -1.41 \\ 2.30 \\ -1.75 \end{bmatrix} \\[2mm]
-0.02x_1(k) - 0.32x_2(k) - 0.02 & \text{if} \begin{bmatrix} 0.26 & 0.96 \\ -0.06 & -1.00 \\ -0.37 & 0.93 \end{bmatrix} x(k) \leq \begin{bmatrix} -1.06 \\ 3.20 \\ -2.30 \end{bmatrix} \\[2mm]
-0.43x_1(k) - 1.80x_2(k) - 1.65 & \text{if} \begin{bmatrix} 0.39 & 0.92 \\ -0.26 & -0.96 \\ -0.16 & 0.97 \end{bmatrix} x(k) \leq \begin{bmatrix} 0.38 \\ 1.06 \\ -1.98 \end{bmatrix} \\[2mm]
0.16x_1(k) - 0.41x_2(k) - 2.21 & \text{if} \begin{bmatrix} 1.00 & 0 \\ -0.37 & 0.93 \\ -0.39 & -0.92 \end{bmatrix} x(k) \leq \begin{bmatrix} 10.00 \\ -2.73 \\ -0.38 \end{bmatrix} \\[2mm]
-0.56x_1(k) - 0.98x_2(k) & \text{if} \begin{bmatrix} 0.16 & -0.99 \\ -0.16 & 0.99 \\ -0.12 & -0.99 \\ 0.12 & 0.99 \\ -0.50 & -0.87 \\ 0.50 & 0.87 \end{bmatrix} x(k) \leq \begin{bmatrix} 1.97 \\ 1.97 \\ 1.41 \\ 1.41 \\ 0.89 \\ 0.89 \end{bmatrix}
\end{cases}
$$

Fig. 4.12 Explicit interpolating control law and explicit MPC control law as piecewise affine functions of state for Example 4.2

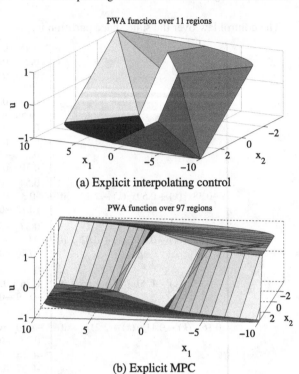

(a) Explicit interpolating control

(b) Explicit MPC

In view of comparison, consider the explicit MPC solution in Example 4.1, Fig. 4.11(a) presents the state space partition of the explicit MPC with the same setup parameters as in Example 4.1. Merging the polyhedral regions with an identical piecewise affine control function, the reduced state space partition is obtained in Fig. 4.11(b).

The comparison of explicit interpolating control and explicit MPC in terms of the number of regions before and after merging is given in Table 4.2.

Figure 4.12 shows the explicit interpolating control law and the explicit MPC control law as piecewise affine functions of state, respectively.

4.4 Improved Interpolating Control

The interpolating controller in Sect. 4.2 and Sect. 4.3 can be considered as an approximate model predictive control law, which in the last decade has received significant attention in the control community [18, 60, 63, 78, 108, 114]. From this point of view, it is worthwhile to obtain an interpolating controller with some given level of accuracy in terms of performance compared with the optimal MPC one. Naturally, the approximation error can be a measure of the level of accuracy. The methods of computing bounds on the approximation error are known, see e.g. [18, 60, 114].

Obviously, the simplest way of improving the performance of the interpolating controller is to use an intermediate s-step controlled invariant set C_s with $1 \leq s < N$. Then there will be not only one level of interpolation but *two* or virtually *any* number of interpolation as necessary from the performance point of view. For simplicity, we provide in the following a study of the case when only one intermediate controlled invariant set C_s is used. Let C_s be in the form,

$$C_s = \{x \in \mathbb{R}^n : F_s x \leq g_s\} \tag{4.32}$$

and satisfy the condition $\Omega_{max} \subset C_s \subset C_N$.

Remark 4.7 It has to be noted however that, the expected increase in performance comes at the price of complexity as long as the intermediate set needs to be stored along with its vertex controller.

For further use, the vertex control law applied for the set C_s is denoted as u_s. Using the same philosophy as in Sect. 4.2, the state x is decomposed as,

1. If $x \in C_N$ and $x \notin C_s$, then

$$x = c_1 x_v + (1 - c_1) x_s \tag{4.33}$$

with $x_v \in C_N$, $x_s \in C_s$ and $0 \leq c_1 \leq 1$. The control law is,

$$u = c_1 u_v + (1 - c_1) u_s \tag{4.34}$$

2. Else $x \in C_s$,

$$x = c_2 x_s + (1 - c_2) x_o \tag{4.35}$$

with $x_s \in C_s$, $x_o \in \Omega_{max}$ and $0 \leq c_2 \leq 1$. The control law is,

$$u = c_2 u_s + (1 - c_2) u_o \tag{4.36}$$

Depending on the value of x, at each time instant, either c_1 or c_2 is minimized in order to be as close as possible to the optimal controller. This can be done by solving the following nonlinear optimization problems,

1. If $x \in C_N \setminus C_s$,

$$c_1^* = \min_{x_v, x_s, c_1} \{c_1\} \tag{4.37}$$

subject to

$$\begin{cases} F_N x_v \leq g_N, \\ F_s x_s \leq g_s, \\ c_1 x_v + (1 - c_1) x_s = x, \\ 0 \leq c_1 \leq 1 \end{cases}$$

2. Else $x \in C_s$,

$$c_2^* = \min_{x_s, x_o, c_2} \{c_2\} \tag{4.38}$$

subject to

$$\begin{cases} F_s x_s \leq g_s, \\ F_o x_o \leq g_o, \\ c_2 x_s + (1 - c_2) x_o = x, \\ 0 \leq c_2 \leq 1 \end{cases}$$

or by changing variables $r_v = c_1 x_v$ and $r_s = c_2 x_s$, the nonlinear optimization problems (4.37) and (4.38) can be transformed in the following LP problems, respectively,

---

1. If $x \in C_N \setminus C_s$

$$c_1^* = \min_{r_v, c_1} \{c_1\} \tag{4.39}$$

subject to

$$\begin{cases} F_N r_v \leq c_1 g_N, \\ F_s (x - r_v) \leq (1 - c_1) g_s, \\ 0 \leq c_1 \leq 1 \end{cases}$$

2. Else $x \in C_s$

$$c_2^* = \min_{r_s, c_2} \{c_2\} \tag{4.40}$$

subject to

$$\begin{cases} F_s r_s \leq c_2 g_s, \\ F_o (x - r_s) \leq (1 - c_2) g_o, \\ 0 \leq c_2 \leq 1 \end{cases}$$

The following theorem shows recursive feasibility and asymptotic stability of the interpolating controller (4.33), (4.34), (4.35), (4.36), (4.39), (4.40),

Theorem 4.8 *The control law* (4.33), (4.34), (4.35), (4.36), (4.39), (4.40) *guarantees recursive feasibility and asymptotic stability of the closed loop system for all initial states* $x(0) \in C_N$.

Proof The proof is omitted here, since it follows the same steps as those presented in the feasibility proof of Theorem 4.1 and the stability proof of Theorem 4.2 in Sect. 4.2.   □

Remark 4.8 Clearly, instead of the second level of interpolation (4.35), (4.36), (4.40), the MPC approach can be applied for all states inside the set C_s. This has very practical consequences in applications, since it is well known [34, 88] that the main issue of MPC for time-invariant linear discrete-time systems is the trade-off between the overall complexity (computational cost) and the size of the domain of attraction. If the prediction horizon is short then the domain of attraction is small. If the prediction horizon is long then the computational cost may be very burdensome for the available hardware. Here MPC with the short prediction horizon is employed inside C_s for the performance and then for enlarging the domain of attraction, the control law (4.33), (4.34), (4.39) is used. In this way one can achieve the performance and the domain of attraction with a relatively small computational cost.

Theorem 4.9 *The control law (4.33), (4.34), (4.35), (4.36), (4.39), (4.40) can be represented as a continuous function of the state.*

Proof Clearly, the discontinuity of the control law may arise only on the boundary of the set C_s, denoted as ∂C_s. Note that for $x \in \partial C_s$, the LP problems (4.39), (4.40) have the trivial solution,

$$c_1^* = 0, \qquad c_2^* = 1$$

Therefore, for $x \in \partial C_s$ the control law (4.33), (4.34), (4.39) is $u = u_s$ and the control law (4.35), (4.36), (4.40) is $u = u_s$. Hence the continuity of the control law is guaranteed. □

Remark 4.9 It is interesting to note that by using $N - 1$ intermediate sets C_i together with the sets C_N and Ω_{max}, a continuous minimum-time controller is obtained, i.e. a controller that steers all state $x \in C_N$ into Ω_{max} in no more than N steps.

Concerning the explicit solution of the control law (4.33), (4.34), (4.35), (4.36), (4.39), (4.40), with the same argument as in Sect. 4.3, it can be concluded that,

- If $x \in C_N \setminus C_s$ (or $x \in C_s \setminus \Omega_{max}$), the smallest value c_1 (or c_2) will be reached when the region $C_N \setminus C_s$ (or $C_s \setminus \Omega_{max}$) is decomposed into polyhedral partitions in form of simplices with vertices both on ∂C_N and on ∂C_s (or on ∂C_s and on $\partial \Omega_{max}$). The control law in each simplex is a piecewise affine function of the state, whose gains are obtained by interpolation of control values at the vertices of the simplex.
- If $x \in \Omega_{max}$, then the control law is the optimal unconstrained controller.

Example 4.3 Consider again Example 4.1. Here one intermediate set C_s with $s = 4$ is introduced. The set of vertices V_s of C_s is,

$$V_s = \begin{bmatrix} 10.00 & -5.95 & -7.71 & -10.00 & -10.00 & 5.95 & 7.71 & 10.00 \\ -0.06 & 2.72 & 2.86 & 1.78 & 0.06 & -2.72 & -2.86 & -1.78 \end{bmatrix}$$
$$(4.41)$$

Fig. 4.13 Two-level interpolation for improving the performance

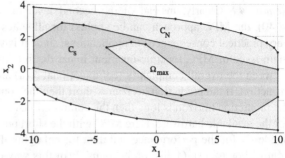

Fig. 4.14 State space partition before merging (number of regions: $N_r = 37$) and after merging ($N_r = 19$), and state trajectories for Example 4.3

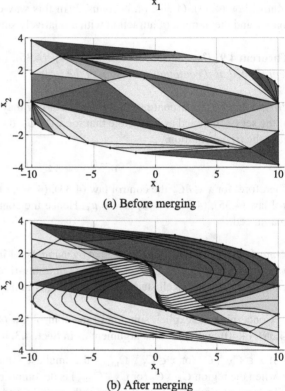

(a) Before merging

(b) After merging

and the set of the corresponding control actions at the vertices V_s is,

$$U_s = \begin{bmatrix} -1 & -1 & -1 & -1 & 1 & 1 & 1 & 1 \end{bmatrix} \tag{4.42}$$

The sets C_N, C_s and $\Omega_{\max}$ are depicted in Fig. 4.13. For the explicit solution, the state space partition of the control law (4.33), (4.34), (4.35), (4.36), (4.39), (4.40) is shown in Fig. 4.14(a). Merging the regions with identical control laws, the reduced state space partition is obtained in Fig. 4.14(b). This figure also shows state trajectories of the closed-loop system for different initial conditions.

Figure 4.15 shows the control law with two-level interpolation.

Fig. 4.15 Control value as a
piecewise affine function of
the state using two-level
interpolation for Example 4.3

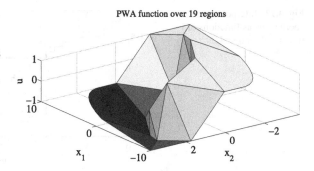

Fig. 4.16 State and input
trajectories for one-level
interpolating control
(*dashed*), and for two-level
interpolating control (*solid*)
for Example 4.3

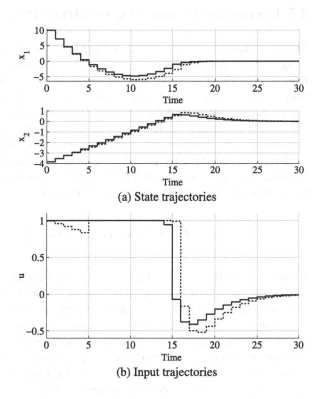

(a) State trajectories

(b) Input trajectories

For the initial condition $x(0) = [9.9800 \ -3.8291]^T$, Fig. 4.16 shows the results of a time-domain simulation. The two curves correspond to the one-level and two-level interpolating control, respectively.

Figure 4.17 presents the interpolating coefficients c_1^* and c_2^*. As expected c_1^* and c_2^* are positive and non-increasing. It is also interesting to note that $\forall k \geq 10$, $c_1^*(k) = 0$, indicating that x is inside C_s and $\forall k \geq 14$, $c_2^*(k) = 0$, indicating that x is inside $\Omega_{\max}$.

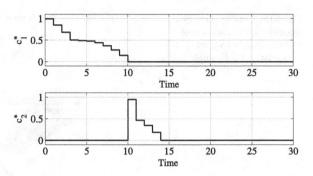

4.5 Interpolating Control via Quadratic Programming

The interpolating controller in Sect. 4.2 and Sect. 4.4 makes use of linear programming, which is extremely simple. However, the main issue regarding the implementation of Algorithm 4.1 is the non-uniqueness of the solution. Multiple optima are undesirable, as they might lead to a fast switching between the different optimal control actions when the LP problem (4.10) is solved on-line. In addition, MPC traditionally has been formulated using a quadratic criterion [92]. Hence, also in interpolating control it is worthwhile to investigate the use of quadratic programming.

Before introducing a QP formulation, let us note that the idea of using QP for interpolating control is not new. In [10, 110], Lyapunov theory is used to compute an upper bound of the infinite horizon cost function,

$$J = \sum_{k=0}^{\infty} \left\{ x(k)^T Q x(k) + u(k)^T R u(k) \right\} \tag{4.43}$$

where $Q \succeq 0$ and $R \succ 0$ are the state and input weighting matrices. At each time instant, the algorithms in [110] use an on-line decomposition of the current state, with each component lying in a separate invariant set, after which the corresponding controller is applied to each component separately in order to calculate the control action. Polytopes are employed as candidate invariant sets. Hence, the on-line optimization problem can be formulated as a QP problem. The approach taken in this section follows ideas originally proposed in [10, 110]. In this setting we provide a QP based solution to the constrained control problem.

This section begins with a brief summary on the works [10, 110]. For this purpose, it is assumed that a set of unconstrained asymptotically stabilizing feedback controllers $u(k) = K_i x(k)$, $i = 1, 2, \ldots, s$ is available such that the corresponding invariant set $\Omega_i \subseteq X$

$$\Omega_i = \left\{ x \in \mathbb{R}^n : F_o^{(i)} x \leq g_o^{(i)} \right\} \tag{4.44}$$

is non-empty for $i = 1, 2, \ldots, s$.

Denote Ω as the convex hull of Ω_i, $i = 1, 2, \ldots, s$. It follows that $\Omega \subseteq X$, since $\Omega_i \subseteq X$, $\forall i = 1, 2, \ldots, s$ and the fact that X is convex. Any state $x(k) \in \Omega$ can be decomposed as,

$$x(k) = \lambda_1(k)\widehat{x}_1(k) + \lambda_2(k)\widehat{x}_2(k) + \cdots + \lambda_s(k)\widehat{x}_s(k) \tag{4.45}$$

where $\widehat{x}_i(k) \in \Omega_i$, $\forall i = 1, 2, \ldots, s$ and $\sum_{i=1}^{s} \lambda_i(k) = 1$, $\lambda_i(k) \geq 0$.

Define $r_i = \lambda_i \widehat{x}_i$. Since $\widehat{x}_i \in \Omega_i$, it follows that $r_i \in \lambda_i \Omega_i$ or equivalently,

$$F_o^{(i)} r_i \leq \lambda_i g_o^{(i)}, \quad \forall i = 1, 2, \ldots, s \tag{4.46}$$

From (4.45), one obtains

$$x(k) = r_1(k) + r_2(k) + \cdots + r_s(k) \tag{4.47}$$

Consider the following control law,

$$u(k) = \sum_{i=1}^{s} \lambda_i K_i \widehat{x}_i = \sum_{i=1}^{s} K_i r_i \tag{4.48}$$

where $u_i(k) = K_i r_i(k)$ is the control law in Ω_i. One has,

$$x(k+1) = Ax(k) + Bu(k) = A \sum_{i=1}^{s} r_i(k) + B \sum_{i=1}^{s} K_i r_i(k) = \sum_{i=1}^{s} (A + BK_i) r_i(k)$$

or,

$$x(k+1) = \sum_{i=1}^{s} r_i(k+1) \tag{4.49}$$

where $r_i(k+1) = A_{ci} r_i(k)$ and $A_{ci} = A + BK_i$.

Define the vector $z \in \mathbb{R}^{sn}$ as,

$$z = \begin{bmatrix} r_1^T & r_2^T & \cdots & r_s^T \end{bmatrix}^T \tag{4.50}$$

Using (4.49), one obtains,

$$z(k+1) = \Phi z(k) \tag{4.51}$$

where

$$\Phi = \begin{bmatrix} A_{c1} & 0 & \cdots & 0 \\ 0 & A_{c2} & \cdots & 0 \\ \vdots & \vdots & \ddots & \vdots \\ 0 & 0 & \cdots & A_{cs} \end{bmatrix}$$

For the given state and control weighting matrices $Q \in \mathbb{R}^{n \times n}$ and $R \in \mathbb{R}^{m \times m}$, consider the following quadratic function,

$$V(z) = z^T P z \tag{4.52}$$

where matrix $P \in \mathbb{R}^{sn \times sn}$, $P \succ 0$ is chosen to satisfy,

$$V(z(k+1)) - V(z(k)) \leq -x(k)^T Q x(k) - u(k)^T R u(k) \qquad (4.53)$$

Using (4.51), the left hand side of (4.53) can be rewritten as,

$$V(z(k+1)) - V(z(k)) = z(k)^T (\Phi^T P \Phi - P) z(k) \qquad (4.54)$$

and using (4.47), (4.48), (4.50), the right hand side of (4.53) becomes,

$$-x(k)^T Q x(k) - u(k)^T R u(k) = z(k)^T (Q_1 + R_1) z(k) \qquad (4.55)$$

where

$$Q_1 = -\begin{bmatrix} I \\ I \\ \vdots \\ I \end{bmatrix} Q \begin{bmatrix} I & I & \cdots & I \end{bmatrix}, \qquad R_1 = -\begin{bmatrix} K_1^T \\ K_2^T \\ \vdots \\ K_s^T \end{bmatrix} R \begin{bmatrix} K_1 & K_2 & \cdots & K_s \end{bmatrix}$$

Combining (4.53), (4.54) and (4.55), one gets,

$$\Phi^T P \Phi - P \preceq Q_1 + R_1$$

or by using the Schur complements, one obtains,

$$\begin{bmatrix} P + Q_1 + R_1 & \Phi^T P \\ P \Phi & P \end{bmatrix} \succeq 0 \qquad (4.56)$$

Problem (4.56) is linear with respect to matrix P. Since matrix Φ has a sub-unitary spectral radius (4.51), problem (4.56) is always feasible. One way to obtain P is to solve the following LMI problem,

$$\min_P \{ \mathrm{trace}(P) \} \qquad (4.57)$$

subject to constraints (4.56).

At each time instant, for a given current state x, consider the following optimization problem,

$$\min_{r_i, \lambda_i} \left\{ \begin{bmatrix} r_1^T & r_2^T & \cdots & r_s^T \end{bmatrix} P \begin{bmatrix} r_1 \\ r_2 \\ \vdots \\ r_s \end{bmatrix} \right\} \qquad (4.58)$$

subject to

$$
\begin{cases}
F_o^{(i)} r_i \leq \lambda_i g_o^{(i)}, & \forall i = 1, 2, \ldots, s, \\[2mm]
\displaystyle\sum_{i=1}^{s} r_i = x, \\[2mm]
\displaystyle\sum_{i=1}^{s} \lambda_i = 1, \\[2mm]
\lambda_i \geq 0, & \forall i = 1, 2, \ldots, s
\end{cases}
$$

and implement as input the control action $u = \sum_{i=1}^{s} K_i r_i$.

Theorem 4.10 [10, 110] *The control law (4.45), (4.48), (4.58) guarantees recursive feasibility and asymptotic stability for all initial states $x(0) \in \Omega$.*

Note that using the approach in [10, 110], for a given state x we are trying to minimize $r_1, r_2, \ldots, r_s$ in the weighted Euclidean norm sense. This is somehow a conflicting task, since,

$$r_1 + r_2 + \cdots + r_s = x$$

In addition, if the first controller is optimal and plays the role of a performance controller, then one would like to have a control law as close as possible to the first controller. This means that in the interpolation scheme (4.45), one would like to have $r_1 = x$ and

$$r_2 = r_3 = \cdots = r_s = 0$$

whenever it is possible. And it is not trivial to do this with the approach in [10, 110].

Below we provide a contribution to this line of research by considering one of the interpolation factors, i.e. control gains to be a performance related one, while the remaining factors play the role of degrees of freedom to enlarge the domain of attraction. This alternative approach can provide the appropriate framework for the constrained control design which builds on the unconstrained optimal controller (generally with high gain) and subsequently need to adjusted them to cope with the constraints and limitations (via interpolation with adequate low gain controllers). From this point of view, in the remaining part of this section we try to build a bridge between the linear interpolation scheme presented in Sect. 4.2 and the QP based interpolation approaches in [10, 110].

For a given set of state and control weighting matrices $Q_i \succeq 0$, $R_i \succ 0$, consider the following set of quadratic functions,

$$V_i(r_i) = r_i^T P_i r_i, \quad \forall i = 2, 3, \ldots, s \tag{4.59}$$

where matrix $P_i \in \mathbb{R}^{n \times n}$ and $P_i \succ 0$ is chosen to satisfy

$$V_i\big(r_i(k+1)\big) - V_i\big(r_i(k)\big) \leq -r_i(k)^T Q_i r_i(k) - u_i(k)^T R_i u_i(k) \tag{4.60}$$

Since $r_i(k+1) = A_{ci}r_i(k)$ and $u_i(k) = K_i r_i(k)$, equation (4.60) can be written as,

$$A_{ci}^T P_i A_{ci} - P_i \preceq -Q_i - K_i^T R_i K_i$$

By using the Schur complements, one obtains

$$\begin{bmatrix} P_i - Q_i - K_i^T R_i K_i & A_{ci}^T P_i \\ P_i A_{ci} & P_i \end{bmatrix} \succeq 0 \qquad (4.61)$$

Since matrix A_{ci} has a sub-unitary spectral radius, problem (4.61) is always feasible. One way to obtain matrix P_i is to solve the following LMI problem,

$$\min_{P_i}\{\text{trace}(P_i)\} \qquad (4.62)$$

subject to constraint (4.61).

Define the vector $z_1 \in \mathbb{R}^{(s-1)(n+1)}$ as,

$$z_1 = \begin{bmatrix} r_2^T & r_3^T & \cdots & r_s^T & \lambda_2 & \lambda_3 & \cdots & \lambda_s \end{bmatrix}^T$$

Consider the following quadratic function,

$$J(z_1) = \sum_{i=2}^{s} r_i^T P_i r_i + \sum_{i=2}^{s} \lambda_i^2 \qquad (4.63)$$

We underline the fact that the sum is built on indices $\{2, 3, \ldots, s\}$, corresponding to the more poorly performing controllers. At each time instant, consider the following optimization problem,

$$V_1(z_1) = \min_{z_1}\{J(z_1)\} \qquad (4.64)$$

subject to the constraints

$$\begin{cases} F_o^{(i)} r_i \leq \lambda_i g_o^{(i)}, \forall i = 1, 2, \ldots, s, \\ \sum_{i=1}^{s} r_i = x, \\ \sum_{i=1}^{s} \lambda_i = 1, \\ \lambda_i \geq 0, \forall i = 1, 2, \ldots, s \end{cases}$$

and apply as input the control signal $u = \sum_{i=1}^{s}\{K_i r_i\}$.

Theorem 4.11 *The control law* (4.45), (4.48), (4.64) *guarantees recursive feasibility and asymptotic stability for all initial states* $x(0) \in \Omega$.

Proof Theorem 4.11 makes two important claims, namely the recursive feasibility and the asymptotic stability. These can be treated sequentially.

Recursive feasibility: It has to be proved that $F_u u(k) \leq g_u$ and $x(k+1) \in \Omega$ for all $x(k) \in \Omega$. It holds that,

$$F_u u(k) = F_u \sum_{i=1}^{s} \lambda_i K_i \widehat{x}_i = \sum_{i=1}^{s} \lambda_i F_u K_i \widehat{x}_i \leq \sum_{i=1}^{s} \lambda_i g_u = g_u$$

and

$$x(k+1) = Ax(k) + Bu(k) = \sum_{i=1}^{s} \lambda_i A_{ci} \widehat{x}_i(k)$$

Since $A_{ci}\widehat{x}_i(k) \in \Omega_i \subseteq \Omega$, it follows that $x(k+1) \in \Omega$.

Asymptotic stability: Consider the positive function $V_1(z_1)$ as a candidate Lyapunov function. From the recursive feasibility proof, it is apparent that if $\lambda_1^*(k)$, $\lambda_2^*(k), \ldots, \lambda_s^*(k)$ and $r_1^*(k), r_2^*(k), \ldots, r_s^*(k)$ is the solution of the optimization problem (4.64) at time instant k, then $\lambda_i(k+1) = \lambda_i^*(k)$ and

$$r_i(k+1) = A_{ci} r_i^*(k)$$

$\forall i = 1, 2, \ldots, s$ is a feasible solution to (4.64) at time instant $k+1$. Since at each time instant we are trying to minimize $J(z_1)$, it follows that

$$V_1\big(z_1^*(k+1)\big) \leq J\big(z_1(k+1)\big)$$

and therefore

$$V_1\big(z_1^*(k+1)\big) - V_1\big(z_1^*(k)\big) \leq J\big(z_1(k+1)\big) - V_1\big(z_1^*(k)\big)$$

together with (4.60), one obtains

$$V_1\big(z_1^*(k+1)\big) - V_1\big(z_1^*(k)\big) \leq -\sum_{i=2}^{s}\big(r_i^T Q_i r_i + u_i^T R_i u_i\big)$$

Hence $V_1(z_1)$ is a Lyapunov function and the control law (4.45), (4.48), (4.64) assures asymptotic stability for all $x \in \Omega$. $\qquad\square$

The constraints of the problem (4.64) can be rewritten as,

$$
\begin{cases}
F_o^{(1)}(x - r_2 - \cdots - r_s) \leq (1 - \lambda_2 - \cdots - \lambda_s)g_o^{(1)} \\
F_o^{(2)}r_2 \leq \lambda_2 g_o^{(2)} \\
\quad \vdots \\
F_o^{(s)}r_s \leq \lambda_s g_o^{(s)} \\
\lambda_i \geq 0, \quad \forall i = 2, \ldots, s \\
\lambda_2 + \lambda_3 + \cdots + \lambda_s \leq 1
\end{cases}
$$

or, equivalently

$$
Gz_1 \leq S + Ex \tag{4.65}
$$

where

$$
G = \begin{bmatrix}
-F_o^{(1)} & -F_o^{(1)} & \cdots & -F_o^{(1)} & g_o^{(1)} & g_o^{(1)} & \cdots & g_o^{(1)} \\
F_o^{(2)} & 0 & \cdots & 0 & -g_o^{(2)} & 0 & \cdots & 0 \\
0 & F_o^{(3)} & \cdots & 0 & 0 & -g_o^{(3)} & \cdots & 0 \\
\vdots & \vdots & \ddots & \vdots & \vdots & \vdots & \ddots & \vdots \\
0 & 0 & \cdots & F_o^{(s)} & 0 & 0 & \cdots & -g_o^{(s)} \\
0 & 0 & \cdots & 0 & -1 & 0 & \cdots & 0 \\
0 & 0 & \cdots & 0 & 0 & -1 & \cdots & 0 \\
\vdots & \vdots & \ddots & \vdots & \vdots & \vdots & \ddots & \vdots \\
0 & 0 & \cdots & 0 & 0 & 0 & \cdots & -1 \\
0 & 0 & \cdots & 0 & 1 & 1 & \cdots & 1
\end{bmatrix},
$$

$$
S = \begin{bmatrix} (g_o^{(1)})^T & 0 & 0 & \cdots & 0 & 0 & 0 & \cdots & 0 & 1 \end{bmatrix}^T
$$

$$
E = \begin{bmatrix} -(F_o^{(1)})^T & 0 & 0 & \cdots & 0 & 0 & 0 & \cdots & 0 & 0 \end{bmatrix}^T
$$

And the objective function (4.64) can be written as,

$$
\min_{z_1}\{z_1^T H z_1\} \tag{4.66}
$$

Algorithm 4.3 Interpolating control via quadratic programming

---

1. Measure the current state $x(k)$.
2. Solve the QP problem (4.66), (4.65).
3. Apply the control input (4.48).
4. Wait for the next time instant $k := k + 1$.
5. Go to step 1 and repeat.

---

where

$$
H = \begin{bmatrix}
P_2 & 0 & \cdots & 0 & 0 & 0 & \cdots & 0 \\
0 & P_3 & \cdots & 0 & 0 & 0 & \cdots & 0 \\
\vdots & \vdots & \ddots & \vdots & \vdots & \vdots & \ddots & \vdots \\
0 & 0 & \cdots & P_s & 0 & 0 & \cdots & 0 \\
0 & 0 & \cdots & 0 & 1 & 0 & \cdots & 0 \\
0 & 0 & \cdots & 0 & 0 & 1 & \cdots & 0 \\
\vdots & \vdots & \ddots & \vdots & \vdots & \vdots & \ddots & \vdots \\
0 & 0 & \cdots & 0 & 0 & 0 & \cdots & 1
\end{bmatrix}
$$

Hence, the optimization problem (4.64) is transformed into the quadratic programming problem (4.66), (4.65).

It is worth noticing that for all $x \in \Omega_1$, the QP problem (4.66), (4.65) has the trivial solution, namely

$$
\begin{cases} r_i = 0, \\ \lambda_i = 0 \end{cases} \quad \forall i = 2, 3, \dots, s
$$

Hence $r_1 = x$ and $\lambda_1 = 1$. That means, inside the set Ω_1, the interpolating controller (4.45), (4.48), (4.64) becomes the optimal unconstrained controller.

Remark 4.10 Note that Algorithm 4.3 requires the solution of the QP problem (4.66) of dimension $(s - 1)(n + 1)$ where s is the number of interpolated controllers and n is the dimension of state. Clearly, solving the QP problem (4.66) can be computationally expensive when the number of interpolated controllers is big. However, it is usually enough with $s = 2$ or $s = 3$ in terms of performance and in terms of the size of the domain of attraction.

Example 4.4 Consider again the system in Example 4.2 with the same state and control constraints. Two linear feedback controllers are chosen as,

$$
\begin{cases} K_1 = [-0.0942 \quad -0.7724] \\ K_2 = [-0.0669 \quad -0.2875] \end{cases} \tag{4.67}
$$

The first controller $u(k) = K_1 x(k)$ is an optimal controller and plays the role of the performance controller, and the second controller $u(k) = K_2 x(k)$ is used to enlarge the domain of attraction.

Figure 4.18(a) shows the invariant sets Ω_1 and Ω_2 correspond to the controllers K_1 and K_2, respectively. Figure 4.18(b) shows state trajectories obtained by solving the QP problem (4.66), (4.65) for different initial conditions.

The sets Ω_1 and Ω_2 are presented in minimal normalized half-space representation as,

$$\Omega_1 = \left\{ x \in \mathbb{R}^2 : \begin{bmatrix} 1.0000 & 0 \\ -1.0000 & 0 \\ -0.1211 & -0.9926 \\ 0.1211 & 0.9926 \end{bmatrix} x \le \begin{bmatrix} 10.0000 \\ 10.0000 \\ 1.2851 \\ 1.2851 \end{bmatrix} \right\}$$

$$\Omega_2 = \left\{ x \in \mathbb{R}^2 : \begin{bmatrix} 1.0000 & 0 \\ -1.0000 & 0 \\ -0.2266 & -0.9740 \\ 0.2266 & 0.9740 \\ 0.7948 & 0.6069 \\ -0.7948 & -0.6069 \\ -0.1796 & -0.9837 \\ 0.1796 & 0.9837 \\ -0.1425 & -0.9898 \\ 0.1425 & 0.9898 \\ -0.1117 & -0.9937 \\ 0.1117 & 0.9937 \\ -0.0850 & -0.9964 \\ 0.0850 & 0.9964 \\ -0.0610 & -0.9981 \\ 0.0610 & 0.9981 \\ -0.0386 & -0.9993 \\ 0.0386 & 0.9993 \\ -0.0170 & -0.9999 \\ 0.0170 & 0.9999 \end{bmatrix} x \le \begin{bmatrix} 10.0000 \\ 10.0000 \\ 3.3878 \\ 3.3878 \\ 8.5177 \\ 8.5177 \\ 3.1696 \\ 3.1696 \\ 3.0552 \\ 3.0552 \\ 3.0182 \\ 3.0182 \\ 3.0449 \\ 3.0449 \\ 3.1299 \\ 3.1299 \\ 3.2732 \\ 3.2732 \\ 3.4795 \\ 3.4795 \end{bmatrix} \right\}$$

For the weighting matrices $Q_2 = I$, $R_2 = 1$, and by solving the LMI problem (4.62), one obtains,

$$P_2 = \begin{bmatrix} 5.1917 & 9.9813 \\ 9.9813 & 101.2651 \end{bmatrix} \tag{4.68}$$

For the initial condition $x(0) = [6.8200 \ 1.8890]^T$, Fig. 4.19(a) and 4.19(b) present the state and input trajectories of the closed loop system for our approach (solid), and for the approach in [110] (dashed).

For [110], the matrix P in the problem (4.57) is computed as,

$$P = \begin{bmatrix} 4.8126 & 2.9389 & 4.5577 & 13.8988 \\ 2.9389 & 7.0130 & 2.2637 & 20.4391 \\ 4.5577 & 2.2637 & 5.1917 & 9.9813 \\ 13.8988 & 20.4391 & 9.9813 & 101.2651 \end{bmatrix}$$

Fig. 4.18 Feasible invariant sets and state trajectories of the closed loop system for Example 4.4

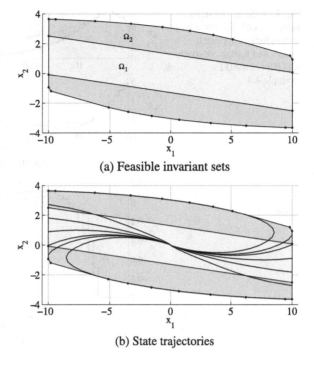

(a) Feasible invariant sets

(b) State trajectories

for the weighting matrices $Q = I$, $R = 1$.

The interpolating coefficient λ_2^* and the Lyapunov function $V_1(z_1)$ are depicted in Fig. 4.20. As expected $V_1(z_1)$ is a positive and non-increasing function.

4.6 Interpolating Control Based on Saturated Controllers

In this section, in order to fully utilize the capability of actuators and to enlarge the domain of attraction, an interpolation between several saturated controllers will be proposed. For simplicity, only single-input single-output system is considered, although extensions to multi-input multi-output systems are straightforward.

From Lemma 2.1 in Sect. 2.4.1, recall that for a given stabilizing controller $u(k) = Kx(k)$, there exists an auxiliary stabilizing controller $u(k) = Hx(k)$ such that the saturation function can be expressed as, $\forall x$ such that $Hx \in U$,

$$\text{sat}\big(Kx(k)\big) = \alpha(k)Kx(k) + \big(1 - \alpha(k)\big)Hx(k) \tag{4.69}$$

where $0 \le \alpha(k) \le 1$. Matrix $H \in \mathbb{R}^n$ can be computed using Theorem 2.3. Using Procedure 2.5 in Sect. 2.4.1, the polyhedral set Ω_s^H can be computed, which is invariant for system,

$$x(k + 1) = Ax(k) + B \, \text{sat}\big(Kx(k)\big) \tag{4.70}$$

and with respect to the constraints (4.2).

Fig. 4.19 State and input trajectories of the closed loop system for our approach (*solid*), and for the approach in [110] (*dashed*) for Example 4.4

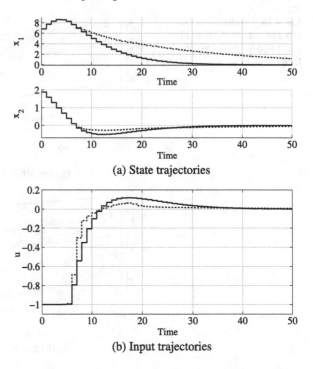

(a) State trajectories

(b) Input trajectories

It is assumed that a set of asymptotically stabilizing feedback controllers $K_i \in \mathbb{R}^n$, $i = 1, 2, \ldots, s$ is available as well as a set of auxiliary matrices $H_i \in \mathbb{R}^n$, $i = 2, \ldots, s$ such that the corresponding invariant sets $\Omega_1 \subseteq X$

$$\Omega_1 = \left\{ x \in \mathbb{R}^n : F_o^{(1)} x \le g_o^{(1)} \right\} \tag{4.71}$$

for the linear controller $u = K_1 x$ and $\Omega_s^{H_i} \subseteq X$

$$\Omega_s^{H_i} = \left\{ x \in \mathbb{R}^n : F_o^{(i)} x \le g_o^{(i)} \right\} \tag{4.72}$$

for the saturated controllers $u = \text{sat}(K_i x)$, $\forall i = 2, 3, \ldots, s$, are non-empty. Denote Ω_s as the convex hull of the sets Ω_1 and $\Omega_s^{H_i}$, $i = 2, 3, \ldots, s$. It follows that $\Omega_s \subseteq X$, since $\Omega_1 \subseteq X$, $\Omega_s^{H_i} \subseteq X$, $\forall i = 2, 3, \ldots, s$ and the fact that X is a convex set.

Remark 4.11 We use one linear control law here in order to show that interpolation can be done between any kind of controllers: *linear or saturated*. The main requirement is that there exists for each of these controllers its own convex invariant set as the domain of attraction.

Any state $x(k) \in \Omega_s$ can be decomposed as,

$$x(k) = \lambda_1(k)\widehat{x}_1(k) + \sum_{i=2}^{s} \lambda_i(k)\widehat{x}_i(k) \tag{4.73}$$

Fig. 4.20 Interpolating coefficient λ_2^* and the Lyapunov function $V_1(z_1)$ as functions of time for Example 4.4

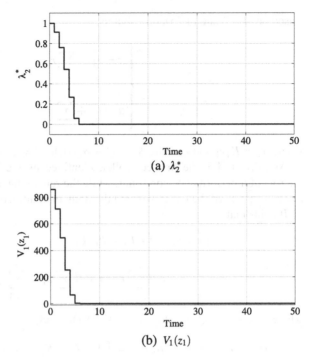

(a) λ_2^*

(b) $V_1(z_1)$

where $\widehat{x}_1(k) \in \Omega_1$, $\widehat{x}_i(k) \in \Omega_s^{H_i}$, $i = 2, 3, \ldots, s$ and

$$\sum_{i=1}^{s} \lambda_i(k) = 1, \quad \lambda_i(k) \geq 0.$$

Consider the following control law,

$$u(k) = \lambda_1(k) K_1 \widehat{x}_1(k) + \sum_{i=2}^{s} \lambda_i(k) \operatorname{sat}\left(K_i \widehat{x}_i(k)\right) \tag{4.74}$$

Using Lemma 2.1, one obtains,

$$u(k) = \lambda_1(k) K_1 \widehat{x}_1(k) + \sum_{i=2}^{s} \lambda_i(k) \left(\alpha_i(k) K_i + \left(1 - \alpha_i(k)\right) H_i\right) \widehat{x}_i(k) \tag{4.75}$$

where $0 \leq \alpha_i(k) \leq 1$ for all $i = 2, 3, \ldots, s$.

Similar with the notation employed in Sect. 4.5, we denote $r_i = \lambda_i \widehat{x}_i$. Since $\widehat{x}_1 \in \Omega_1$ and $\widehat{x}_i \in \Omega_s^{H_i}$, it follows that $r_1 \in \lambda_1 \Omega_1$ and $r_i \in \lambda_i \Omega_s^{H_i}$ or, equivalently

$$F_o^{(i)} r_i \leq \lambda_i g_o^{(i)}, \quad \forall i = 1, 2, \ldots, s \tag{4.76}$$

Based on (4.73) and (4.75), one obtains,

$$
\begin{cases}
x = r_1 + \displaystyle\sum_{i=2}^{s} r_i, \\[4mm]
u = u_1 + \displaystyle\sum_{i=2}^{s} u_i
\end{cases}
\tag{4.77}
$$

where $u_1 = K_1 r_1$ and $u_i = (\alpha_i K_i + (1 - \alpha_i) H_i) r_i$, $i = 2, 3, \ldots, s$.

As in Sect. 4.5, the first controller, identified by the high gain K_1, will play the role of a performance controller, while the remaining controllers $u = \mathrm{sat}(K_i x)$, $i = 2, 3, \ldots, s$ will be used to extend the domain of attraction.

It holds that,

$$
x(k+1) = Ax(k) + Bu(k)
$$

$$
= A \sum_{i=1}^{s} r_i(k) + B \sum_{i=1}^{s} u_i = \sum_{i=1}^{s} r_i(k+1)
$$

where $r_1(k+1) = Ar_1 + Bu_1 = (A + BK_1)r_1$ and

$$
r_i(k+1) = Ar_i(k) + Bu_i(k) = \left\{ A + B\big(\alpha_i K_i + (1 - \alpha_i) H_i\big) \right\} r_i(k)
\tag{4.78}
$$

or, equivalently

$$
r_i(k+1) = A_{ci} r_i(k)
\tag{4.79}
$$

with $A_{ci} = A + B(\alpha_i K_i + (1 - \alpha_i) H_i)$, $\forall i = 2, 3, \ldots, s$.

For a given set of state and control weighting matrices $Q_i \succeq 0$ and $R_i \succ 0$, $i = 2, 3, \ldots, s$, consider the following set of quadratic functions,

$$
V_i(r_i) = r_i^T P_i r_i, \quad i = 2, 3, \ldots, s
\tag{4.80}
$$

where the matrix $P_i \in \mathbb{R}^{n \times n}$, $P_i \succ 0$ is chosen to satisfy,

$$
V_i\big(r_i(k+1)\big) - V_i\big(r_i(k)\big) \le -r_i(k)^T Q_i r_i(k) - u_i(k)^T R_i u_i(k)
\tag{4.81}
$$

With the same argument as in Sect. 4.5, equation (4.81) can be rewritten as,

$$
A_{ci}^T P_i A_{ci} - P_i \preceq -Q_i - \big(\alpha_i K_i + (1 - \alpha_i) H_i\big)^T R_i \big(\alpha_i K_i + (1 - \alpha_i) H_i\big)
$$

Using the Schur complements, the above condition can be transformed into,

$$
\begin{bmatrix} P_i - Q_i - Y_i^T R_i Y_i & A_{ci}^T P_i \\ P_i A_{ci} & P_i \end{bmatrix} \succeq 0
$$

where $Y_i = \alpha_i K_i + (1 - \alpha_i) H_i$. Or, equivalently

$$
\begin{bmatrix} P_i & A_{ci}^T P_i \\ P_i A_{ci} & P_i \end{bmatrix} - \begin{bmatrix} Q_i + Y_i^T R_i Y_i & 0 \\ 0 & 0 \end{bmatrix} \succeq 0
$$

Denote $\sqrt{Q_i}$ and $\sqrt{R_i}$ as the Cholesky factor of the matrices Q_i and R_i, which satisfy

$$\sqrt{Q_i}^T \sqrt{Q_i} = Q_i \quad \text{and} \quad \sqrt{R_i}^T \sqrt{R_i} = R_i.$$

The previous condition can be rewritten as,

$$\begin{bmatrix} P_i & A_{ci}^T P_i \\ P_i A_{ci} & P_i \end{bmatrix} - \begin{bmatrix} \sqrt{Q_i}^T & Y_i^T \sqrt{R_i}^T \\ 0 & 0 \end{bmatrix} \begin{bmatrix} \sqrt{Q_i} & 0 \\ \sqrt{R_i} Y_i & 0 \end{bmatrix} \succeq 0$$

or by using the Schur complements, one obtains,

$$\begin{bmatrix} P_i & A_{ci}^T P_i & \sqrt{Q_i}^T & Y_i^T \sqrt{R_i}^T \\ P_i A_{ci} & P_i & 0 & 0 \\ \sqrt{Q_i} & 0 & I & 0 \\ \sqrt{R_i} Y_i & 0 & 0 & I \end{bmatrix} \succeq 0 \qquad (4.82)$$

Since $Y_i = \alpha_i K_i + (1 - \alpha_i) H_i$, and $A_{ci} = A + B Y_i$ the left hand side of inequality (4.82) is linear in α_i, and hence reaches its minimum at either $\alpha_i = 0$ or $\alpha_i = 1$. Consequently, the set of LMI conditions to be checked is following,

$$\begin{cases} \begin{bmatrix} P_i & (A + B K_i)^T P_i & \sqrt{Q_i}^T & (\sqrt{R_i} K_i)^T \\ P_i (A + B K_i) & P_i & 0 & 0 \\ \sqrt{Q_i} & 0 & I & 0 \\ \sqrt{R_i} K_i & 0 & 0 & I \end{bmatrix} \succeq 0 \\ \begin{bmatrix} P_i & (A + B H_i)^T P_i & \sqrt{Q_i}^T & (\sqrt{R_i} H_i)^T \\ P_i (A + B H_i) & P_i & 0 & 0 \\ \sqrt{Q_i} & 0 & I & 0 \\ \sqrt{R_i} H_i & 0 & 0 & I \end{bmatrix} \succeq 0 \end{cases} \qquad (4.83)$$

Condition (4.83) is linear with respect to the matrix P_i. One way to calculate P_i is to solve the following LMI problem,

$$\min_{P_i} \{ \text{trace}(P_i) \} \qquad (4.84)$$

subject to constraint (4.83).

Once the matrices P_i, $i = 2, 3, \ldots, s$ are computed, they can be used in practice for real-time control based on the following algorithm, which can be formulated as an optimization problem that is efficient with respect to structure and complexity. At each time instant, for a given current state x, minimize on-line the quadratic cost function,

$$\min_{r_i, \lambda_i} \left\{ \sum_{i=2}^{s} r_i^T P_i r_i + \sum_{i=2}^{s} \lambda_i^2 \right\} \tag{4.85}$$

subject to the linear constraints

$$\begin{cases} F_o^{(i)} r_i \leq \lambda_i g_o^{(i)}, & \forall i = 1, 2, \ldots, s, \\ \sum_{i=1}^{s} r_i = x, \\ \sum_{i=1}^{s} \lambda_i = 1, \\ \lambda_i \geq 0, & \forall i = 1, 2, \ldots, s \end{cases}$$

Theorem 4.12 *The control law* (4.73), (4.74), (4.85) *guarantees recursive feasibility and asymptotic stability of the closed loop system for all initial states* $x(0) \in \Omega_s$.

Proof The proof is similar to Theorem 4.11. Hence it is omitted here.   □

Example 4.5 Consider again the system in Example 4.1 with the same state and control constraints. Two gain matrices are chosen as,

$$\begin{cases} K_1 = [-0.9500 \quad -1.1137], \\ K_2 = [-0.4230 \quad -2.0607] \end{cases} \tag{4.86}$$

Using Theorem 2.3, matrix H_2 is computed as,

$$H_2 = [-0.0669 \quad -0.2875] \tag{4.87}$$

The invariant sets Ω_1 and $\Omega_s^{H_2}$ are, respectively constructed for the controllers $u = K_1 x$ and $u = \text{sat}(K_2 x)$, see Fig. 4.21(a). Figure 4.21(b) shows state trajectories for different initial conditions.

The sets Ω_1 and $\Omega_s^{H_2}$ are presented in minimal normalized half-space representation as,

$$\Omega_1 = \left\{ x \in \mathbb{R}^2 : \begin{bmatrix} 0.3919 & -0.9200 \\ -0.3919 & 0.9200 \\ -0.6490 & -0.7608 \\ 0.6490 & 0.7608 \end{bmatrix} x \leq \begin{bmatrix} 1.4521 \\ 1.4521 \\ 0.6831 \\ 0.6831 \end{bmatrix} \right\}$$

Fig. 4.21 Feasible invariant sets and state trajectories of the closed loop system for Example 4.5

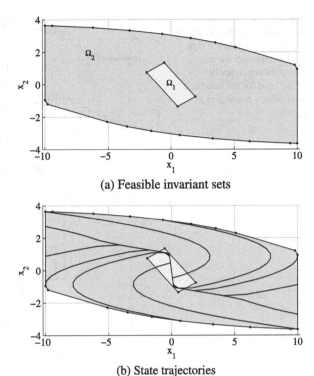

(a) Feasible invariant sets

(b) State trajectories

$$\Omega_s^{H_2} = \left\{ x \in \mathbb{R}^2 : \begin{bmatrix} -0.0170 & -0.9999 \\ 0.0170 & 0.9999 \\ -0.0386 & -0.9993 \\ 0.0386 & 0.9993 \\ -0.0610 & -0.9981 \\ 0.0610 & 0.9981 \\ -0.0850 & -0.9964 \\ 0.0850 & 0.9964 \\ -0.1117 & -0.9937 \\ 0.1117 & 0.9937 \\ -0.1425 & -0.9898 \\ 0.1425 & 0.9898 \\ 0.7948 & 0.6069 \\ -0.7948 & -0.6069 \\ -0.1796 & -0.9837 \\ 0.1796 & 0.9837 \\ 1.0000 & 0 \\ -1.0000 & 0 \\ -0.2266 & -0.9740 \\ 0.2266 & 0.9740 \end{bmatrix} x \le \begin{bmatrix} 3.4795 \\ 3.4795 \\ 3.2732 \\ 3.2732 \\ 3.1299 \\ 3.1299 \\ 3.0449 \\ 3.0449 \\ 3.0182 \\ 3.0182 \\ 3.0552 \\ 3.0552 \\ 8.5177 \\ 8.5177 \\ 3.1696 \\ 3.1696 \\ 10.0000 \\ 10.0000 \\ 3.3878 \\ 3.3878 \end{bmatrix} \right\}$$

Fig. 4.22 State and input trajectories of the closed loop system as functions of time for Example 4.5 for the interpolating controller (*solid*) and for the saturated controller $u = \text{sat}(K_2 x)$ (*dashed*)

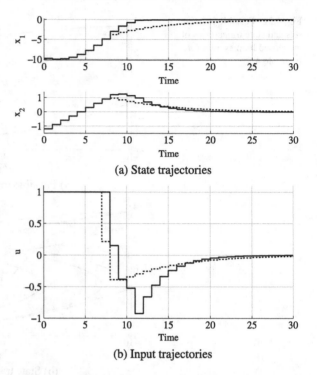

(a) State trajectories

(b) Input trajectories

With the weighting matrices $Q_2 = I$, $R_2 = 0.001$ and by solving the LMI problem (4.84), one obtains,

$$P_2 = \begin{bmatrix} 5.4929 & 9.8907 \\ 9.8907 & 104.1516 \end{bmatrix}$$

For the initial condition $x(0) = [-9.79 \ -1.2]^T$, Fig. 4.22 presents the state and input trajectories for the interpolating controller (solid blue) and for the saturated controller $u = \text{sat}(K_2 x)$ (dashed red), which is the controller corresponding to the set $\Omega_s^{H_2}$. The interpolating coefficient λ_2^* and the objective function as a Lyapunov function are shown in Fig. 4.23.

4.7 Convex Hull of Ellipsoids

For high dimensional systems, the polyhedral based interpolation approaches in Sects. 4.2, 4.3, 4.4, 4.5, 4.6 might be impractical due to the huge number of vertices or half-spaces in the representation of polyhedral sets. In that case, ellipsoids might be a suitable class of sets for interpolation.

Note that the idea of using ellipsoids for a constrained control system is well known, for time-invariant linear continuous-time systems, see [56], and for time-invariant linear discrete-time systems, see [10]. In these papers, a method to construct a continuous control law based on a set of *linear* control laws was proposed

Fig. 4.23 Interpolating coefficient λ_2^* and Lyapunov function as functions of time for Example 4.5

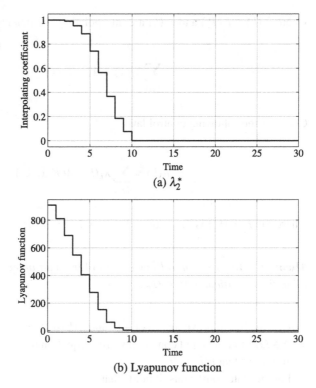

(a) λ_2^*

(b) Lyapunov function

to make the convex hull of an associated set of invariant ellipsoids *invariant*. However these results do not allow to impose priority among the control laws.

In this section, an interpolation using a set of *saturated* controllers and its associated set of invariant ellipsoid is presented. The main contribution with respect to [10, 56] is to provide a new type of controller, that uses interpolation.

It is assumed that a set of asymptotically stabilizing saturated controllers $u = \text{sat}(K_i x)$ is available such that the corresponding ellipsoidal invariant sets $E(P_i)$

$$E(P_i) = \left\{ x \in \mathbb{R}^n : x^T P_i^{-1} x \leq 1 \right\} \qquad (4.88)$$

are non-empty for $i = 1, 2, \ldots, s$. Recall that for all $x(k) \in E(P_i)$, it follows that $\text{sat}(K_i x) \in U$ and $x(k+1) = Ax(k) + B \, \text{sat}(K_i x(k)) \in X$. Denote $\Omega_E \subset \mathbb{R}^n$ as the convex hull of $E(P_i)$, $i = 1, 2, \ldots, s$. It follows that $\Omega_E \subseteq X$, since X is convex and $E(P_i) \subseteq X$, $i = 1, 2, \ldots, s$.

Any state $x(k) \in \Omega_E$ can be decomposed as,

$$x(k) = \sum_{i=1}^{s} \lambda_i(k) \widehat{x}_i(k) \qquad (4.89)$$

where $\widehat{x}_i(k) \in E(P_i)$ and $\lambda_i(k)$ are interpolating coefficients, that satisfy

$$\sum_{i=1}^{s} \lambda_i(k) = 1, \quad \lambda_i(k) \geq 0$$

Consider the following control law,

$$u(k) = \sum_{i=1}^{s} \lambda_i(k) \, \mathrm{sat}\big(K_i \widehat{x}_i(k)\big) \tag{4.90}$$

where $\mathrm{sat}(K_i \widehat{x}_i(k))$ is the saturated control law in $E(P_i)$.

Theorem 4.13 *The control law* (4.89), (4.90) *guarantees recursive feasibility for all initial conditions* $x(0) \in \Omega_E$.

Proof One has to prove that $u(k) \in U$ and $x(k+1) = Ax(k) + Bu(k) \in \Omega_E$ for all $x(k) \in \Omega_E$. For the input constraints, from equation (4.90) and since $\mathrm{sat}(K_i \widehat{x}_i(k)) \in U$, it follows that $u(k) \in U$.

For the state constraints, it holds that,

$$x(k+1) = Ax(k) + Bu(k)$$

$$= A \sum_{i=1}^{s} \lambda_i(k) \widehat{x}_i(k) + B \sum_{i=1}^{s} \lambda_i(k) \, \mathrm{sat}(K_i \widehat{x}_i(k))$$

$$= \sum_{i=1}^{s} \lambda_i(k)(A \widehat{x}_i(k) + B \, \mathrm{sat}(K_i \widehat{x}_i(k)))$$

One has $A \widehat{x}_i(k) + B \, \mathrm{sat}(K_i \widehat{x}_i(k)) \in E(P_i) \subseteq \Omega_E$, $i = 1, 2, \ldots, s$, which ultimately assures that $x(k+1) \in \Omega_E$. □

As in Sects. 4.5 and 4.6, the first high gain controller will be used for the performance, while the rest of available low gain controllers will be used to enlarge the domain of attraction. For a given current state x, consider the following optimization problem,

$$\lambda_i^* = \min_{\widehat{x}_i, \lambda_i} \left\{ \sum_{i=2}^{s} \lambda_i \right\} \tag{4.91}$$

subject to

$$\begin{cases} \widehat{x}_i^T P_i^{-1} \widehat{x}_i \leq 1, & \forall i = 1, 2, \ldots, s, \\[2mm] \displaystyle\sum_{i=1}^{s} \lambda_i \widehat{x}_i = x, \\[2mm] \displaystyle\sum_{i=1}^{s} \lambda_i = 1, \\[2mm] \lambda_i \geq 0, & \forall i = 1, 2, \ldots, s \end{cases}$$

Theorem 4.14 *The control law* (4.89), (4.90), (4.91) *guarantees asymptotic stability for all initial states* $x(0) \in \Omega_E$.

Proof Consider the following non-negative function,

$$V(x) = \sum_{i=2}^{s} \lambda_i^*(k) \tag{4.92}$$

for all $x \in \Omega_E \setminus E(P_1)$. $V(x)$ is a Lyapunov function candidate.

For any $x(k) \in \Omega_E \setminus E(P_1)$, by solving the optimization problem (4.91) and by applying (4.89), (4.90), one obtains

$$\begin{cases} x(k) = \displaystyle\sum_{i=1}^{s} \lambda_i^*(k) \widehat{x}_i^*(k) \\[4mm] u(k) = \displaystyle\sum_{i=1}^{s} \lambda_i^*(k) \, \mathrm{sat}(K_i \widehat{x}_i^*(k)) \end{cases}$$

It follows that,

$$x(k+1) = Ax(k) + Bu(k) = A \sum_{i=1}^{s} \lambda_i^*(k) \widehat{x}_i^*(k) + B \sum_{i=1}^{s} \lambda_i^*(k) \, \mathrm{sat}(K_i \widehat{x}_i^*(k))$$

$$= \sum_{i=1}^{s} \lambda_i^*(k) \widehat{x}_i(k+1)$$

where $\widehat{x}_i(k+1) = A\widehat{x}_i^*(k) + B \, \mathrm{sat}(K_i \widehat{x}_i^*(k)) \in E(P_i)$, $\forall i = 1, 2, \ldots, s$. Hence $\lambda_i^*(k)$, $\forall i = 1, 2, \ldots, s$ is a feasible solution of (4.91) at time $k + 1$.

At time $k + 1$, by soling the optimization problem (4.91), one obtains

$$x(k+1) = \sum_{i=1}^{s} \lambda_i^*(k+1) \widehat{x}_i^*(k+1)$$

where $\widehat{x}_i^*(k+1) \in E(P_i)$. It follows that $\sum_{i=2}^{s} \lambda_i^*(k+1) \leq \sum_{i=2}^{s} \lambda_i^*(k)$ and $V(x)$ is a non-increasing function.

The contractive property of the ellipsoids $E(P_i)$, $i = 1, 2, \ldots, s$ assures that there is no initial condition $x(0) \in \Omega_E \setminus E(P_1)$ such that $\sum_{i=2}^{s} \lambda_i^*(k+1) = \sum_{i=2}^{s} \lambda_i^*(k)$ for sufficiently large and finite k. It follows that $V(x) = \sum_{i=2}^{s} \lambda_i^*(k)$ is a Lyapunov function for all $x \in \Omega_E \setminus E(P_1)$.

The proof is completed by noting that inside $E(P_1)$, $\lambda_1 = 1$ and $\lambda_i = 0$, $i = 2, 3, \ldots, s$, the saturated controller $u = \text{sat}(K_1\widehat{x})$ is contractive and thus the control laws (4.89), (4.90), (4.91) assures asymptotic stability for all $x \in \Omega_E$.   □

Denote $r_i = \lambda_i \widehat{x}_i$. Since $\widehat{x}_i \in E(P_i)$, it follows that $r_i \in \lambda_i E(P_i)$, and hence $r_i^T P_i^{-1} r_i \leq \lambda_i^2$. The non-linear optimization problem (4.91) can be rewritten as,

$$\min_{r_i, \lambda_i} \left\{ \sum_{i=2}^{s} \lambda_i \right\} \tag{4.93}$$

subject to

$$\begin{cases} r_i^T P_i^{-1} r_i \leq \lambda_i^2, \quad \forall i = 1, 2, \ldots, s, \\ \sum_{i=1}^{s} r_i = x, \\ \sum_{i=1}^{s} \lambda_i = 1, \quad \lambda_i \geq 0, \quad \forall i = 1, 2, \ldots, s \end{cases}$$

By using the Schur complements, (4.93) is converted into the following LMI problem,

$$\min_{r_i, \lambda_i} \left\{ \sum_{i=2}^{s} \lambda_i \right\} \tag{4.94}$$

subject to

$$\begin{cases} \begin{bmatrix} \lambda_i & r_i^T \\ r_i & \lambda_i P_i \end{bmatrix} \succeq 0, \quad \forall i = 1, 2, \ldots, s, \\ \sum_{i=1}^{s} r_i = x, \\ \sum_{i=1}^{s} \lambda_i = 1, \ \lambda_i \geq 0, \quad \forall i = 1, 2, \ldots, s \end{cases}$$

Algorithm 4.4 Interpolating control—Convex hull of ellipsoids
1. Measure the current state $x(k)$.
2. Solve the LMI problem (4.94).
3. Apply as input the control signal (4.90).
4. Wait for the next time instant $k := k + 1$.
5. Go to step 1 and repeat.

Remark 4.12 It is worth noticing that for all $x(k) \in E(P_1)$, the LMI problem (4.94) has the trivial solution,

$$\lambda_i = 0, \quad \forall i = 2, 3, \ldots, s$$

Hence $\lambda_1 = 1$ and $x = \widehat{x}_1$. In this case, the interpolating controller becomes the saturated controller $u = \text{sat}(K_1 x)$.

Example 4.6 Consider again the system in Example 4.1 with the same state and control constraints. Three gain matrices are chosen as,

$$\begin{cases} K_1 = [-0.9500 \quad -1.1137], \\ K_2 = [-0.4230 \quad -2.0607], \\ K_3 = [-0.5010 \quad -2.1340] \end{cases} \tag{4.95}$$

Fig. 4.24 Invariant ellipsoids and state trajectories of the closed loop system for Example 4.6

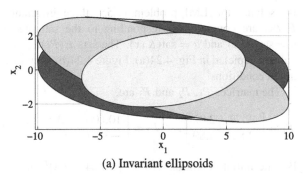

(a) Invariant ellipsoids

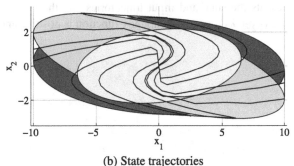

(b) State trajectories

Fig. 4.25 State trajectory, input trajectory and the sum $(\lambda_2^* + \lambda_3^*)$ of the closed loop system for Example 4.6

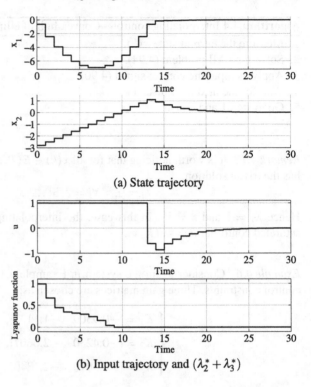

(a) State trajectory

(b) Input trajectory and $(\lambda_2^* + \lambda_3^*)$

By solving the LMI problem (2.55) three invariant ellipsoids $E(P_1)$, $E(P_2)$, $E(P_3)$ are computed corresponding to the saturated controllers $u = \text{sat}(K_1 x)$, $u = \text{sat}(K_2 x)$ and $u = \text{sat}(K_3 x)$. The sets $E(P_1)$, $E(P_2)$, $E(P_3)$ and their convex hull are depicted in Fig. 4.24(a). Figure 4.24(b) shows state trajectories for different initial conditions.

The matrices P_1, P_2 and P_3 are,

$$P_1 = \begin{bmatrix} 42.27 & 2.82 \\ 2.82 & 4.80 \end{bmatrix}, \quad P_2 = \begin{bmatrix} 100.00 & -3.10 \\ -3.10 & 8.12 \end{bmatrix}, \quad P_3 = \begin{bmatrix} 100.00 & -19.40 \\ -19.40 & 9.54 \end{bmatrix}$$

For the initial condition $x(0) = [-0.64 \ -2.8]^T$, using Algorithm 4.4, Fig. 4.25 presents the state and input trajectories and the sum $(\lambda_2^* + \lambda_3^*)$. As expected, the sum $(\lambda_2^* + \lambda_3^*)$, i.e. the Lyapunov function is positive and non-increasing.

Chapter 5
Interpolating Control—Robust State Feedback Case

5.1 Problem Formulation

Consider the problem of regulating to the origin the following uncertain and/or time-varying linear discrete-time system subject to additive bounded disturbances,

$$x(k+1) = A(k)x(k) + B(k)u(k) + Dw(k) \tag{5.1}$$

where $x(k) \in \mathbb{R}^n$, $u(k) \in \mathbb{R}^m$ and $w(k) \in \mathbb{R}^d$ are respectively, the measurable state, the input and the disturbance vectors. The matrices $A(k) \in \mathbb{R}^{n \times n}$, $B(k) \in \mathbb{R}^{n \times m}$ and $D \in \mathbb{R}^{n \times d}$. $A(k)$ and $B(k)$ satisfy,

$$
\begin{cases}
A(k) = \displaystyle\sum_{i=1}^{q} \alpha_i(k) A_i, \qquad B(k) = \displaystyle\sum_{i=1}^{q} \alpha_i(k) B_i, \\[2mm]
\displaystyle\sum_{i=1}^{q} \alpha_i(k) = 1, \quad \alpha_i(k) \geq 0
\end{cases}
\tag{5.2}
$$

where the matrices A_i, B_i are given.

The state, the control and the disturbance are subject to the following bounded polytopic constraints,

$$
\begin{cases}
x(k) \in X, & X = \{ x \in \mathbb{R}^n : F_x x \leq g_x \} \\
u(k) \in U, & U = \{ u \in \mathbb{R}^m : F_u u \leq g_u \} \\
w(k) \in W, & W = \{ w \in \mathbb{R}^d : F_w w \leq g_w \}
\end{cases}
\tag{5.3}
$$

where the matrices F_x, F_u and F_w and the vectors g_x, g_u and g_w are assumed to be constant with $g_x > 0$, $g_u > 0$, $g_w > 0$. The inequalities are component-wise.

H.-N. Nguyen, *Constrained Control of Uncertain, Time-Varying, Discrete-Time Systems*, 115
Lecture Notes in Control and Information Sciences 451,
DOI 10.1007/978-3-319-02827-9_5,
© Springer International Publishing Switzerland 2014

5.2 Interpolating Control via Linear Programming

It is assumed that an unconstrained robustly asymptotically stabilizing feedback controller

$$u(k) = Kx(k)$$

is available such that the corresponding maximal robustly invariant set $\Omega_{max} \subseteq X$,

$$\Omega_{max} = \{x \in \mathbb{R}^n : F_o x \leq g_o\} \tag{5.4}$$

is non-empty. Furthermore with some given and fixed integer $N > 0$, based on Procedure 2.3 the robustly controlled invariant set $C_N \subseteq X$,

$$C_N = \{x \in \mathbb{R}^n : F_N x \leq g_N\} \tag{5.5}$$

is computed such that all $x \in C_N$ can be steered into Ω_{max} in no more than N steps when suitable control is applied. The set C_N is decomposed into a set of simplices $C_N^{(j)}$, each formed by n vertices of C_N and the origin. For all $x \in C_N$, the vertex controller

$$u(k) = K^{(j)}x(k), \quad x \in C_N^{(j)} \tag{5.6}$$

where $K^{(j)}$ is defined as in (3.38) robustly stabilizes the system (5.1), while the constraints (5.3) are fulfilled.

In the robust case, similar to the nominal case presented in Sect. 4.2, Chap. 4, the weakness of vertex control is that the full control range is exploited only on the border of the set C_N in the state space, with progressively smaller control action when state approaches the origin. Hence the time to regulate the plant to the origin is longer than necessary. Here we provide a method to overcome this shortcoming, where the control action is still smooth. For this purpose, any state $x(k) \in C_N$ is decomposed as,

$$x(k) = c(k)x_v(k) + (1 - c(k))x_o(k) \tag{5.7}$$

where $x_v \in C_N$, $x_o \in \Omega_{max}$ and $0 \leq c \leq 1$.

Consider the following control law,

$$u(k) = c(k)u_v(k) + (1 - c(k))u_o(k) \tag{5.8}$$

where $u_v(k)$ is the vertex control law (5.6) for $x_v(k)$ and $u_o(k) = Kx_o(k)$ is the control law in Ω_{max}.

Theorem 5.1 *For system (5.1) and constraints (5.3), the control law (5.7), (5.8) guarantees recursive feasibility for all initial states $x(0) \in C_N$.*

Proof For recursive feasibility, it has to be proved that,

$$\begin{cases} F_u u(k) \le g_u, \\ x(k+1) = A(k)x(k) + B(k)u(k) + Dw(k) \in C_N \end{cases}$$

for all $x(k) \in C_N$.

While the feasibility of the input constraints is proved in a similar way to the nominal case, the state constraint feasibility deserves an adaptation.

$$\begin{aligned} x(k+1) &= A(k)x(k) + B(k)u(k) + Dw(k) \\ &= A(k)\{c(k)x_v(k) + (1 - c(k))x_o(k)\} \\ &\quad + B(k)\{c(k)u_v(k) + (1 - c(k))u_o(k)\} + Dw(k) \\ &= c(k)x_v(k+1) + (1 - c(k))x_o(k+1) \end{aligned}$$

where

$$x_v(k+1) = A(k)x_v(k) + B(k)u_v(k) + Dw(k) \in C_N$$
$$x_o(k+1) = A(k)x_o(k) + B(k)u_o(k) + Dw(k) \in \Omega_{\max} \subseteq C_N$$

It follows that $x(k+1) \in C_N$.   □

As in Sect. 4.2, in order for $u(k)$ in (5.8) to be as close as possible to the optimal unconstrained local controller, one would like to minimize the interpolating coefficient $c(k)$. This can be done by solving the following nonlinear optimization problem,

$$c^* = \min_{x_v, x_o, c} \{c\} \tag{5.9}$$

subject to

$$\begin{cases} F_N x_v \le g_N, \\ F_o x_o \le g_o, \\ cx_v + (1 - c)x_o = x, \\ 0 \le c \le 1 \end{cases}$$

Define $r_v = cx_v$ and $r_o = (1 - c)x_o$. Since $x_v \in C_N$ and $x_o \in \Omega_{\max}$, it follows that $r_v \in cC_N$ and $r_o \in (1 - c)\Omega_{\max}$ or equivalently,

$$\begin{cases} F_N r_v \le cg_N, \\ F_o r_o \le (1 - c)g_o \end{cases}$$

With this change of variables, the nonlinear optimization problem (5.9) is transformed into the following linear programming problem,

$$c^* = \min_{r_v, c}\{c\} \tag{5.10}$$

subject to

$$\begin{cases} F_N r_v \leq c g_N, \\ F_o(x - r_v) \leq (1-c)g_o, \\ 0 \leq c \leq 1 \end{cases}$$

Theorem 5.2 *The control law* (5.7), (5.8), (5.10) *guarantees robustly asymptotic stability*[1] *for all initial states* $x(0) \in C_N$.

Proof First of all we will prove that all solutions starting in $C_N \setminus \Omega_{max}$ will reach Ω_{max} in *finite time*. For this purpose, consider the following non-negative function,

$$V(x) = c^*(x), \quad \forall x \in C_N \setminus \Omega_{max} \tag{5.11}$$

$V(x)$ is a candidate Lyapunov function. At time k, after solving the LP problem (5.10) and applying (5.7), (5.8), one obtains, for $x(k) \in C_N \setminus \Omega_{max}$,

$$\begin{cases} x(k) = c^*(k)x_v^*(k) + \big(1 - c^*(k)\big)x_o^*(k), \\ u(k) = c^*(k)u_v(k) + \big(1 - c^*(k)\big)u_o(k) \end{cases}$$

It follows that,

$$x(k+1) = A(k)x(k) + B(k)u(k) + Dw(k)$$

$$= c^*(k)x_v(k+1) + \big(1 - c^*(k)\big)x_o(k+1)$$

where

$$\begin{cases} x_v(k+1) = A(k)x_v^*(k) + B(k)u_v(k) + Dw(k) \in C_N, \\ x_o(k+1) = A(k)x_o^*(k) + B(k)u_o(k) + Dw(k) \in \Omega_{max} \end{cases}$$

Hence $\{c^*(k), x_v^*(k), x_o^*(k)\}$ is a feasible solution for the LP problem (5.10) at time $k+1$, see Fig. 5.1. By solving (5.10) at time $k+1$, one gets the optimal solution, namely

$$x(k+1) = c^*(k+1)x_v^*(k+1) + \big(1 - c^*(k+1)\big)x_o^*(k+1)$$

---

[1] Here by robustly asymptotic stability we understand that the state of the closed loop system converges to the minimal robustly positively invariant set [97, 105] of the system,

$$x(k+1) = \big(A(k) + B(k)K\big)x(k) + Dw(k)$$

despite the parameter variation and the influence of additive disturbances.

Fig. 5.1 Graphical illustration for the proof of Theorem 5.2

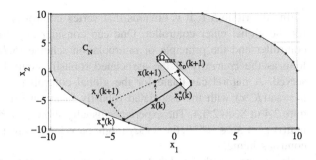

Algorithm 5.1 Interpolating control—Implicit solution

1. Measure the current state $x(k)$.
2. Solve the LP problem (5.10).
3. Compute u_v by determining to which simplex x_v^* belongs and using (5.6).
4. Implement as input the control action (5.8).
5. Wait for the next time instant $k := k + 1$.
6. Go to step 1 and repeat.

---

where $x_v^*(k+1) \in C_N$ and $x_o^*(k+1) \in \Omega_{\max}$. It follows that $c^*(k+1) \leq c^*(k)$ and $V(x)$ is non-increasing.

With the same argument as in the proof of Theorem 4.3, it can be shown that the interpolating coefficient c in (5.10) is minimal if and only if x is decomposed as $x = cx_v + (1-c)x_o$ with $x_v \in \partial C_N$ and $x_o \in \partial \Omega_{\max}$, see Fig. 5.1. Recall here that $\partial(\cdot)$ denotes the boundary of the corresponding set $(\cdot)$.

The state contraction properties of the vertex controller and the local controller for the states on ∂C_N and on $\partial \Omega_{\max}$, respectively, guarantee that there is no initial condition $x(0) \in C_N \setminus \Omega_{\max}$ such that $c^*(k) = c^*(0)$ for sufficiently large and finite k. The conclusion is that $V(x) = c^*(x)$ is a Lyapunov function for $x(k) \in C_N \setminus \Omega_{\max}$ and all solutions starting in $C_N \setminus \Omega_{\max}$ will reach $\Omega_{\max}$ in *finite time*.

The proof is complete by noting that inside $\Omega_{\max}$, the LP problem (5.10) has the trivial solution $c^* = 0$. Hence the control law (5.7), (5.8), (5.10) becomes the local controller, which is robustly stabilizing. Therefore robustly asymptotic stability is guaranteed $\forall x \in C_N$.   □

It is worth noticing that the complexity of the control law (5.7), (5.8), (5.10) is in direct relationship with the complexity of the vertex controller and can be very high, since in general the complexity of the set C_N is high in terms of vertices. It is also well known [26] that the number of simplices of vertex control is typically much greater than the number of vertices. Therefore a question is how to achieve an interpolating controller whose complexity is not correlated with the complexity of the involved sets.

In Algorithm 5.1, it is obvious that vertex control is only one possible choice for the global outer controller. One can consider any other linear or non-linear controller and the principle of interpolation scheme (5.7), (5.8), (5.10) holds as long as the *convexity* of the associated robustly controlled invariant set is preserved. A natural candidate for the global controller is the saturated controller $u = \text{sat}(K_s x)$ with the associated robustly invariant set Ω_s computed by Procedure 2.4 in Sect. 2.3.4. The experience usually shows that by properly choosing the matrix gain $K_s \in \mathbb{R}^{m \times n}$, the set Ω_s might be as large as that of any other constrained control scheme.

In summary with the global saturated controller $u(k) = \text{sat}(K_s x(k))$ the interpolating controller (5.7), (5.8), (5.10) involves the following steps,

---

1. Design a local gain K and a global gain K_s, both stabilizing with some desired performance specifications. Usually K is chosen for the performance, while K_s is designed for extending the domain of attraction.
2. Compute the invariant sets Ω_{max} and Ω_s associated with the controllers $u = Kx$ and $u = \text{sat}(K_s x)$, respectively. Ω_{max} is computed by using Procedure 2.2, and Ω_s by Procedure 2.4.
3. Implement the control law (5.7), (5.8), (5.10).

Practically, the interpolation scheme using saturated control is simpler than the interpolation scheme using vertex control, while the domain of attraction remains typically the same.

Remark 5.1 Concerning the explicit solution of the control law (5.7), (5.8), (5.10) with the vertex controller, using the same argument as in Sect. 4.3, it can be concluded that,

- If $x \in C_N \setminus \Omega_{\text{max}}$, the smallest value of the interpolating coefficient c will be reached when the region $C_N \setminus \Omega$ is decomposed into polyhedral partitions in form of simplices with vertices both on ∂C_N and on $\partial \Omega_{\text{max}}$. The control law in each simplex is a piecewise affine function of state whose gains are obtained by interpolation of control values at the vertices of the simplex.
- If $x \in \Omega_{\text{max}}$, then the control law is the optimal local controller.

Example 5.1 Consider the following uncertain and time-varying linear discrete-time system,

$$x(k+1) = A(k)x(k) + B(k)u(k) \qquad (5.12)$$

where

$$\begin{cases} A(k) = \alpha(k)A_1 + \big(1 - \alpha(k)\big)A_2 \\ B(k) = \alpha(k)B_1 + \big(1 - \alpha(k)\big)B_2 \end{cases}$$

and

$$A_1 = \begin{bmatrix} 1 & 0.1 \\ 0 & 1 \end{bmatrix}, \qquad A_2 = \begin{bmatrix} 1 & 0.2 \\ 0 & 1 \end{bmatrix}, \qquad B_1 = \begin{bmatrix} 0 \\ 1 \end{bmatrix}, \qquad B_2 = \begin{bmatrix} 0 \\ 1.5 \end{bmatrix}$$

At each time instant $\alpha(k) \in [0, 1]$ is an uniformly distributed pseudo-random number. The constraints are,

$$-10 \le x_1 \le 10, \qquad -10 \le x_2 \le 10, \qquad -1 \le u \le 1 \qquad (5.13)$$

The stabilizing feedback gain for states near the origin is chosen as,

$$K = [-1.8112 \quad -0.8092]$$

Using Procedure 2.2 and Procedure 2.3, the sets Ω_{max} and C_N are obtained and shown in Fig. 5.1. Note that $C_{27} = C_{28}$ is the maximal robustly controlled invariant set for system (5.12).

The set of vertices of C_N is given by the matrix $V(C_N)$ below, together with the control matrix U_v at these vertices,

$$V(P_N) = [V_1 \quad -V_1],$$

$$U_v = [U_1 \quad -U_1],$$

$$V_1 = \begin{bmatrix} 10.0000 & 9.7000 & 9.1000 & 8.2000 & 7.0000 & 5.5000 & 3.7000 & 2.3000 & -10.0000 \\ 0 & 1.5000 & 3.0000 & 4.5000 & 6.0000 & 7.5000 & 9.0000 & 10.0000 & 10.0000 \end{bmatrix},$$

$$U_1 = \begin{bmatrix} -1 & -1 & -1 & -1 & -1 & -1 & -1 & -1 & -1 \end{bmatrix}$$

The set Ω_{max} is presented in minimal normalized half-space representation as,

$$\Omega_{max} = \left\{ x \in \mathbb{R}^2 : \begin{bmatrix} -0.9130 & -0.4079 \\ 0.9130 & 0.4079 \\ 0.8985 & -0.4391 \\ -0.8985 & 0.4391 \\ 1.0000 & 0.0036 \\ -1.0000 & -0.0036 \\ 0.9916 & 0.1297 \\ -0.9916 & -0.1297 \end{bmatrix} \le \begin{bmatrix} 0.5041 \\ 0.5041 \\ 2.3202 \\ 2.3202 \\ 1.3699 \\ 1.3699 \\ 1.1001 \\ 1.1001 \end{bmatrix} \right\}$$

Solving explicitly the LP problem (5.10) by using multi-parametric linear programming, the state space partition is obtained in Fig. 5.2(a). The number of regions is $N_r = 27$. Merging the regions with the identical control law, the reduced state space partition is obtained ($N_r = 13$) in Fig. 5.2(b). Figure 5.2(b) also presents state trajectories for different initial conditions and realizations of $\alpha(k)$.

The control law over the state space partition with 13 regions is,

$$
u(k) = \begin{cases}
-1 & \text{if } \begin{bmatrix} 0.93 & 0.37 \\ 0.78 & 0.62 \\ 0.64 & 0.77 \\ -0.91 & -0.41 \\ -0.90 & 0.44 \\ 0.58 & 0.81 \\ 0.71 & 0.71 \\ 0.86 & 0.51 \\ 0.20 & -0.98 \\ 0.98 & 0.20 \end{bmatrix} x(k) \le \begin{bmatrix} 9.56 \\ 9.21 \\ 9.28 \\ -0.50 \\ 2.32 \\ 9.47 \\ 9.19 \\ 9.35 \\ 2.00 \\ 9.81 \end{bmatrix} \\[4pt]
-1 & \text{if } \begin{bmatrix} 0.90 & -0.44 \\ -0.00 & 1.00 \\ -0.59 & -0.81 \end{bmatrix} x(k) \le \begin{bmatrix} -2.32 \\ 10.00 \\ -2.14 \end{bmatrix} \\[4pt]
-0.92 x_1(k) - 1.25 x_2(k) + 2.31 & \text{if } \begin{bmatrix} 0.90 & -0.44 \\ 0.59 & 0.81 \\ -0.66 & -0.75 \end{bmatrix} x(k) \le \begin{bmatrix} -2.32 \\ 2.14 \\ -0.95 \end{bmatrix} \\[4pt]
-0.00 x_1(k) - 0.20 x_2(k) + 1.00 & \text{if } \begin{bmatrix} 0.66 & 0.75 \\ -1.00 & 0.00 \\ 0.27 & -0.96 \end{bmatrix} x(k) \le \begin{bmatrix} 0.95 \\ 10.00 \\ -2.75 \end{bmatrix} \\[4pt]
0.17 x_1(k) - 0.80 x_2(k) + 2.72 & \text{if } \begin{bmatrix} 1.00 & 0.00 \\ 0.23 & -0.97 \\ -0.27 & 0.96 \end{bmatrix} x(k) \le \begin{bmatrix} -1.37 \\ -2.31 \\ 2.75 \end{bmatrix} \\[4pt]
0.11 x_1(k) - 0.56 x_2(k) + 2.13 & \text{if } \begin{bmatrix} 0.99 & 0.13 \\ 0.20 & -0.98 \\ -0.23 & 0.97 \end{bmatrix} x(k) \le \begin{bmatrix} -1.10 \\ -2.00 \\ 2.31 \end{bmatrix} \\[4pt]
-1.81 x_1(k) - 0.81 x_2(k) & \text{if } \begin{bmatrix} -0.91 & -0.41 \\ 0.91 & 0.41 \\ 0.90 & -0.44 \\ -0.90 & 0.44 \\ 1.00 & 0.00 \\ -1.00 & -0.00 \\ 0.99 & 0.13 \\ -0.99 & -0.13 \end{bmatrix} x(k) \le \begin{bmatrix} 0.50 \\ 0.50 \\ 2.32 \\ 2.32 \\ 1.37 \\ 1.37 \\ 1.10 \\ 1.10 \end{bmatrix}
\end{cases}
$$

(due to symmetry of the explicit solution, only the control law for seven regions are reported here)

The interpolating coefficient and the control input are depicted in Fig. 5.3.

For the initial condition $x_0 = [2.2954 \ 9.9800]^T$, Fig. 5.4 shows the state and input trajectories as functions of time for the interpolating controller (solid). As a comparison, Fig. 5.4 also shows the state and input trajectories obtained by using the LMI based MPC algorithm in [74] (dashed). The state and control weighting matrices are $Q = I$, $R = 0.01$ for [74].

Figure 5.5 presents the interpolating coefficient and the realization of $\alpha(k)$. As expected $c^*(k)$ is a positive and non-increasing function.

Fig. 5.2 Explicit solution before and after merging for the interpolating controller and state trajectories of the closed loop system for Example 5.1

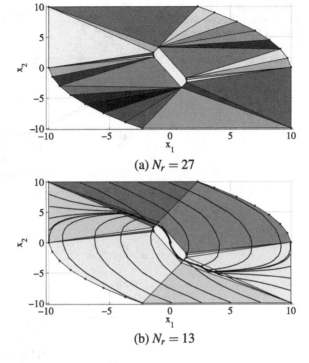

(a) $N_r = 27$

(b) $N_r = 13$

Fig. 5.3 Interpolating coefficient and the control input as piecewise affine functions of state for Example 5.1

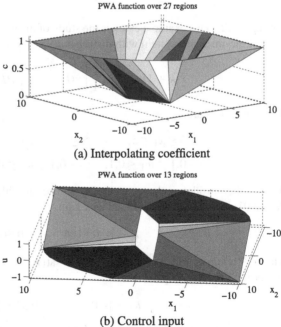

PWA function over 27 regions

(a) Interpolating coefficient

PWA function over 13 regions

(b) Control input

Fig. 5.4 State and input trajectories as functions of time for Example 5.1 for the interpolating controller (*solid*), and for [74] (*dashed*)

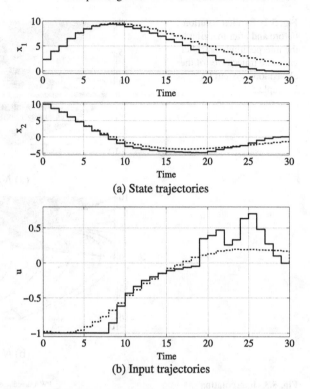

(a) State trajectories

(b) Input trajectories

Example 5.2 This example extends the study of the nominal case. The discrete-time linear time-invariant system with disturbances is given as,

$$x(k+1) = \begin{bmatrix} 1 & 1 \\ 0 & 1 \end{bmatrix} x(k) + \begin{bmatrix} 0 \\ 1 \end{bmatrix} u(k) + w(k) \tag{5.14}$$

The constraints are,

$$-5 \le x_1 \le 5, \qquad -5 \le x_2 \le 5, \qquad -1 \le u \le 1,$$
$$-0.1 \le w_1 \le 0.1, \qquad -0.1 \le w_2 \le 0.1 \tag{5.15}$$

The local controller is chosen as an LQ controller with weighting matrices $Q = I$, $R = 0.01$, leading to the state feedback gain,

$$K = [-0.6136 \quad -1.6099]$$

The following saturated controller $u(k) = \mathrm{sat}(K_s x(k))$ is chosen as a global controller with the matrix gain,

$$K_s = [-0.1782 \quad -0.5205]$$

Using Procedure 2.2 and Procedure 2.4 the sets $\Omega_{\max}$ and Ω_s are respectively, computed for the control laws $u(k) = Kx(k)$ and $u(k) = \mathrm{sat}(K_s x(k))$, see Fig. 5.6(a).

Fig. 5.5 Interpolating coefficient and the realization of $\alpha(k)$ as functions of time for Example 5.1

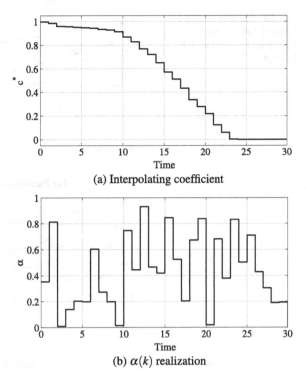

(a) Interpolating coefficient

(b) $\alpha(k)$ realization

Note that Ω_s is actually the maximal invariant set for system (5.14) with constraints (5.15), which can be verified by comparing the equivalence between the set Ω_s and its one-step robustly controlled invariant set. Figure 5.6(b) presents state trajectories for different initial conditions and realizations of $w(k)$. It is worth noticing that the trajectories do not converge to the origin but to the minimal robustly invariant set of the system,

$$x(k+1) = (A + BK)x(k) + w(k)$$

Ω_s and $\Omega_{\max}$ are presented in minimal normalized half-space representation as,

$$\Omega_s = \left\{ x \in \mathbb{R}^2 : \begin{bmatrix} 1.0000 & 0 \\ -1.0000 & 0 \\ 0.7071 & 0.7071 \\ -0.7071 & -0.7071 \\ 0.4472 & 0.8944 \\ -0.4472 & -0.8944 \\ 0.3162 & 0.9487 \\ -0.3162 & -0.9487 \\ 0.2425 & 0.9701 \\ -0.2425 & -0.9701 \\ 0.1961 & 0.9806 \\ -0.1961 & -0.9806 \end{bmatrix} x \leq \begin{bmatrix} 5.0000 \\ 5.0000 \\ 3.4648 \\ 3.4648 \\ 2.5491 \\ 2.5491 \\ 2.3401 \\ 2.3401 \\ 2.4254 \\ 2.4254 \\ 2.6476 \\ 2.6476 \end{bmatrix} \right\}$$

Fig. 5.6 Feasible invariant sets and state trajectories of the closed loop system for Example 5.2

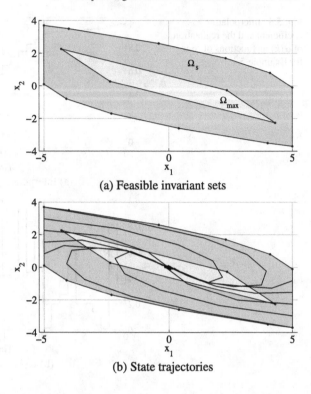

(a) Feasible invariant sets

(b) State trajectories

Fig. 5.7 State trajectories as functions of time for Example 5.2 for the interpolating controller (*solid*), and for the tube MPC in [81] (*dashed*)

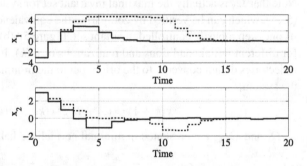

$$\Omega_{\max} = \left\{ x \in \mathbb{R}^2 : \begin{bmatrix} -0.3562 & -0.9344 \\ 0.3562 & 0.9344 \\ 0.7129 & 0.7013 \\ -0.7129 & -0.7013 \end{bmatrix} x \leq \begin{bmatrix} 0.5804 \\ 0.5804 \\ 1.4813 \\ 1.4813 \end{bmatrix} \right\}$$

For the initial condition $x(0) = [-3.0654 \; 2.9541]^T$, Fig. 5.7 and Fig. 5.8 show the state and input trajectories for the interpolating controller (solid). As a comparison, we choose the tube MPC in [81]. The dashed lines in Fig. 5.7 and Fig. 5.8 are obtained by using this technique.

Fig. 5.8 Input trajectories as functions of time for Example 5.2 for the interpolating controller (*solid*), and for the tube MPC in [81] (*dashed*)

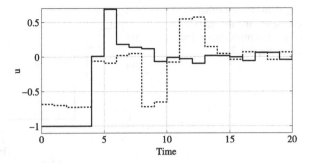

Fig. 5.9 Outer invariant approximation of the minimal robustly invariant set for Example 5.2

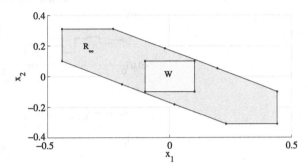

The following parameters were used for the tube MPC. The outer invariant approximation R_∞ of the minimal robustly invariant set was constructed for system,

$$x(k+1) = (A + BK)x(k) + w(k)$$

using the method in [105]. The set R_∞ is depicted in Fig. 5.9. The setup of the MPC approach for the nominal system of the tube MPC framework is $Q = I$, $R = 0.01$. The prediction horizon $N = 10$.

The interpolating coefficient $c^*(k)$ and the realization of $w(k)$ are presented in Fig. 5.10.

5.3 Interpolating Control via Quadratic Programming

The non-uniqueness of the solution is the main issue regarding the implementation of the interpolating controller in Sect. 5.2. Hence, as in the nominal case, it is also worthwhile in the robust case to have an interpolation scheme with strictly convex objective function.

In this section, we consider the problem of regulating to the origin system (5.1) in the absence of disturbances. In other words, the system under consideration is of the form,

$$x(k+1) = A(k)x(k) + B(k)u(k) \tag{5.16}$$

where the uncertainty description of $A(k)$ and $B(k)$ is as in (5.2).

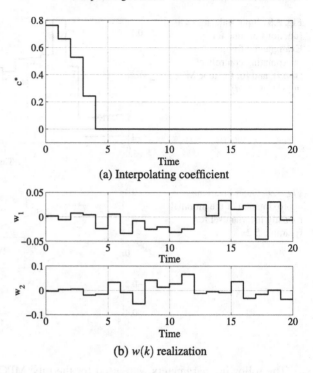

Fig. 5.10 Interpolating coefficient and the realization of $w(k)$ as functions of time for Example 5.2

(a) Interpolating coefficient

(b) $w(k)$ realization

For a given set of robustly asymptotically stabilizing controllers $u(k) = K_i x(k)$, $i = 1, 2, \ldots, s$ and corresponding maximal robustly invariant sets $\Omega_i \subseteq X$

$$\Omega_i = \left\{ x \in \mathbb{R}^n : F_o^{(i)} x \le g_o^{(i)} \right\} \tag{5.17}$$

denote Ω as a convex hull of Ω_i. It follows from the convexity of X that $\Omega \subseteq X$, since $\Omega_i \subseteq X$, $\forall i = 1, 2, \ldots, s$.

By employing the same design scheme in Sect. 4.5, the first high gain controller in this enumeration will play the role of a performance controller, while the remaining low gain controllers will be used in the interpolation scheme to enlarge the domain of attraction. Any state $x(k) \in \Omega$ can be decomposed as follows,

$$x(k) = \lambda_1(k)\widehat{x}_1(k) + \lambda_2(k)\widehat{x}_2(k) + \cdots + \lambda_s(k)\widehat{x}_s(k) \tag{5.18}$$

where $\widehat{x}_i(k) \in \Omega_i$, $\forall i = 1, 2, \ldots, s$ and

$$\sum_{i=1}^{s} \lambda_i(k) = 1, \quad \lambda_i(k) \ge 0$$

Consider the following control law,

$$u(k) = \lambda_1(k)K_1\widehat{x}_1(k) + \lambda_2(k)K_2\widehat{x}_2(k) + \cdots + \lambda_s(k)K_s\widehat{x}_s(k) \tag{5.19}$$

where $u_i(k) = K_i\widehat{x}_i(k)$ is the control law in Ω_i. Define $r_i = \lambda_i\widehat{x}_i$. Since $\widehat{x}_i \in \Omega_i$, it follows that $r_i \in \lambda_i\Omega_i$ or equivalently that the set of inequalities,

$$F_o^{(i)}r_i \le \lambda_i g_o^{(i)} \tag{5.20}$$

is verified $\forall i = 1, 2, \ldots, s$.

It holds that,

$$x(k+1) = A(k)x(k) + B(k)u(k) = A(k)\sum_{i=1}^{s} r_i(k) + B(k)\sum_{i=1}^{s} K_i r_i(k)$$

$$= \sum_{i=1}^{s}(A(k) + B(k)K_i)r_i(k)$$

or

$$x(k+1) = \sum_{i=1}^{s} r_i(k+1) \tag{5.21}$$

with $r_i(k+1) = A_{ci}(k)r_i(k)$ and $A_{ci}(k) = A(k) + B(k)K_i$.

For a given set of state and control weighting matrices $Q_i \succeq 0$, $R_i \succ 0$, $i = 2, 3, \ldots, s$, consider the following set of quadratic functions,

$$V_i(r_i) = r_i^T P_i r_i, \quad \forall i = 2, 3, \ldots, s \tag{5.22}$$

where $P_i \in \mathbb{R}^{n \times n}$ and $P_i \succ 0$ is chosen to satisfy,

$$V_i(r_i(k+1)) - V_i(r_i(k)) \le -r_i(k)^T Q_i r_i(k) - u_i(k)^T R_i u_i(k) \tag{5.23}$$

From (5.22), (5.23) and since $r_i(k+1) = A_{ci}(k)r_i(k)$, $u_i(k) = K_i r_i(k)$, it follows that,

$$A_{ci}^T P_i A_{ci} - P_i \preceq -Q_i - K_i^T R_i K_i$$

By using the Schur complements, one obtains

$$\begin{bmatrix} P_i - Q_i - K_i^T R_i K_i & A_{ci}^T P_i \\ P_i A_{ci} & P_i \end{bmatrix} \succ 0 \tag{5.24}$$

$A_{ci}(k) = A(k) + B(k)K_i = \sum_{j=1}^{q} \alpha_j(k)(A_j + B_j K_i)$ is linear with respect to $\alpha_j(k)$. Hence one should verify (5.24) at the vertices of $\alpha_j(k)$, i.e. when $\alpha_j(k) = 0$ or $\alpha_j(k) = 1$. Therefore the set of LMI conditions to be satisfied is following,

$$\begin{bmatrix} P_i - Q_i - K_i^T R_i K_i & (A_j + B_j K_i)^T P_i \\ P_i(A_j + B_j K_i) & P_i \end{bmatrix} \succ 0, \quad \forall j = 1, 2, \ldots, q \tag{5.25}$$

One way to obtain matrix P_i is to solve the following LMI problem,

$$\min_{P_i}\{\text{trace}(P_i)\} \tag{5.26}$$

subject to constraints (5.25).

Define the vector $z \in \mathbb{R}^{(s-1)(n+1)}$ as follows,

$$z = \begin{bmatrix} r_2^T & \cdots & r_s^T & \lambda_2 & \cdots & \lambda_s \end{bmatrix}^T \tag{5.27}$$

At each time instant, consider the following optimization problem,

---

$$V(z) = \min_z \left\{ \sum_{i=2}^{s} r_i^T P_i r_i + \sum_{i=2}^{s} \lambda_i^2 \right\} \tag{5.28}$$

subject to the constraints

$$\begin{cases} F_o^{(i)} r_i \leq \lambda_i g_o^{(i)}, \ i = 1, 2, \ldots, s, \\[2mm] \sum_{i=1}^{s} r_i = x, \\[2mm] \sum_{i=1}^{s} \lambda_i = 1, \\[2mm] \lambda_i \geq 0, \ i = 1, 2, \ldots, s \end{cases}$$

and apply as input the control action $u = \sum_{i=1}^{s} K_i r_i$.

We underline the fact that the cost function is built on the indices $\{2, 3, \ldots, s\}$, which correspond to the more poorly performing controllers. Simultaneously, the cost function is intended to diminish the influence of these controller actions in the interpolation scheme toward the unconstrained optimum with $r_i = 0$ and $\lambda_i = 0$, $\forall i = 2, 3, \ldots, s$

Theorem 5.3 *The interpolating controller* (5.18), (5.19), (5.28) *guarantees recursive feasibility and robustly asymptotic stability for all initial states* $x(0) \in \Omega$.

Proof The proof of this theorem follows the same argumentation as the one of Theorem 4.11. Hence it is omitted here.   □

As in Sect. 4.5, the cost function in (5.28) can be rewritten in a quadratic form as,

$$V(z) = \min_z \{z^T H z\} \tag{5.29}$$

with

$$H = \begin{bmatrix} P_2 & 0 & \cdots & 0 & 0 & 0 & \cdots & 0 \\ 0 & P_3 & \cdots & 0 & 0 & 0 & \cdots & 0 \\ \vdots & \vdots & \ddots & \vdots & 0 & 0 & \cdots & 0 \\ 0 & 0 & \cdots & P_s & 0 & 0 & \cdots & 0 \\ 0 & 0 & \cdots & 0 & 1 & 0 & \cdots & 0 \\ 0 & 0 & \cdots & 0 & 0 & 1 & \cdots & 0 \\ \vdots & \vdots & \ddots & \vdots & 0 & 0 & \cdots & 0 \\ 0 & 0 & \cdots & 0 & 0 & 0 & \cdots & 1 \end{bmatrix}$$

and the constraints of (5.28) can be rewritten as,

$$Gz \le S + Ex(k) \tag{5.30}$$

where

$$G = \begin{bmatrix} -F_o^{(1)} & -F_o^{(1)} & \cdots & -F_o^{(1)} & g_o^{(1)} & g_o^{(1)} & \cdots & g_o^{(1)} \\ F_o^{(2)} & 0 & \cdots & 0 & -g_o^{(2)} & 0 & \cdots & 0 \\ 0 & F_o^{(3)} & \cdots & 0 & 0 & -g_o^{(3)} & \cdots & 0 \\ \vdots & \vdots & \ddots & \vdots & \vdots & \vdots & \ddots & \vdots \\ 0 & 0 & \cdots & F_o^{(s)} & 0 & 0 & \cdots & -g_o^{(s)} \\ 0 & 0 & \cdots & 0 & -1 & 0 & \cdots & 0 \\ 0 & 0 & \cdots & 0 & 0 & -1 & \cdots & 0 \\ \vdots & \vdots & \ddots & \vdots & \vdots & \vdots & \ddots & \vdots \\ 0 & 0 & \cdots & 0 & 0 & 0 & \cdots & -1 \\ 0 & 0 & \cdots & 0 & 1 & 1 & \cdots & 1 \end{bmatrix}$$

$$S = \begin{bmatrix} (g_o^{(1)})^T & 0 & 0 & \cdots & 0 & 0 & 0 & \cdots & 0 & 1 \end{bmatrix}^T$$

$$E = \begin{bmatrix} -(F_o^{(1)})^T & 0 & 0 & \cdots & 0 & 0 & 0 & \cdots & 0 & 0 \end{bmatrix}^T$$

Hence the optimization problem (5.28) is transformed into the quadratic programming problem (5.29), (5.30).

It is worth noticing that for all $x \in \Omega_1$, the QP problem (5.29), (5.30) has the trivial solution, that is,

$$\begin{cases} r_i = 0, \\ \lambda_i = 0, \end{cases} \quad \forall i = 2, 3, \ldots s$$

Hence $r_1 = x$ and $\lambda_1 = 1$ for $x \in \Omega_1$. And therefore, inside Ω_1, the interpolating controller becomes the optimal one.

Algorithm 5.2 Interpolating control via quadratic programming

---
1. Measure the current state $x(k)$.
2. Solve the QP problem (5.29), (5.30).
3. Implement as input the control action $u = \sum_{i=1}^{s} K_i r_i$.
4. Wait for the next time instant $k := k + 1$.
5. Go to step 1 and repeat.

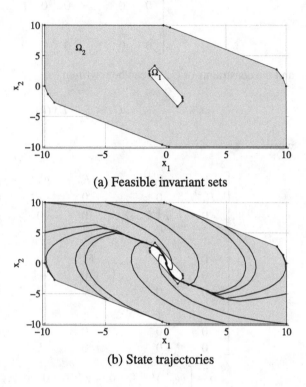

Fig. 5.11 Feasible invariant sets and state trajectories of the closed loop system for Example 5.3

(a) Feasible invariant sets

(b) State trajectories

Example 5.3 Consider the system in Example 5.1 with the same state and control constraints. Two linear feedback controllers are chosen as,

$$\begin{cases} K_1 = [-1.8112 \quad -0.8092], \\ K_2 = [-0.0786 \quad -0.1010] \end{cases} \tag{5.31}$$

The first controller $u(k) = K_1 x(k)$ plays the role of a performance controller, while the second controller $u(k) = K_2 x(k)$ is used for extending the domain of attraction.

Figure 5.11(a) shows the maximal robustly invariant sets Ω_1 and Ω_2 correspond to the controllers K_1 and K_2 respectively. Figure 5.11(b) presents state trajectories for different initial conditions and realizations of $\alpha(k)$.

The sets Ω_1 and Ω_2 are presented in minimal normalized half-space representation as,

$$\Omega_1 = \left\{ x \in \mathbb{R}^2 : \begin{bmatrix} 1.0000 & 0.0036 \\ -1.0000 & -0.0036 \\ 0.9916 & 0.1297 \\ -0.9916 & -0.1297 \\ 0.8985 & -0.4391 \\ -0.8985 & 0.4391 \\ -0.9130 & -0.4079 \\ 0.9130 & 0.4079 \end{bmatrix} x \leq \begin{bmatrix} 1.3699 \\ 1.3699 \\ 1.1001 \\ 1.1001 \\ 2.3202 \\ 2.3202 \\ 0.5041 \\ 0.5041 \end{bmatrix} \right\}$$

$$\Omega_2 = \left\{ x \in \mathbb{R}^2 : \begin{bmatrix} 0.9352 & 0.3541 \\ -0.9352 & -0.3541 \\ 0.9806 & 0.1961 \\ -0.9806 & -0.1961 \\ -0.5494 & -0.8355 \\ 0.5494 & 0.8355 \\ 1.0000 & 0 \\ 0 & 1.0000 \\ -1.0000 & 0 \\ 0 & -1.0000 \\ -0.6142 & -0.7892 \\ 0.6142 & 0.7892 \end{bmatrix} x \leq \begin{bmatrix} 9.5779 \\ 9.5779 \\ 9.8058 \\ 9.8058 \\ 8.2385 \\ 8.2385 \\ 10.0000 \\ 10.0000 \\ 10.0000 \\ 10.0000 \\ 7.8137 \\ 7.8137 \end{bmatrix} \right\}$$

With the weighting matrices $Q_2 = I$, $R_2 = 0.001$, and by solving the LMI problem (5.26), one obtains,

$$P_2 = \begin{bmatrix} 17.5066 & 7.9919 \\ 7.9919 & 16.7525 \end{bmatrix}$$

For the initial condition $x(0) = [0.3380 \ 9.6181]^T$, Fig. 5.12(a) and 5.12(b) show the state and input trajectories for our approach (solid). Figure 5.12(a) and 5.12(b) also show the state and input trajectories for the algorithm in [102] (dashed). The state and control weighting matrices for [102] are $Q = I$, $R = 0.001$.

Figure 5.13 presents the interpolating coefficient $\lambda_2(k)$, the realization of $\alpha(k)$ and the Lyapunov function as functions of time.

5.4 Interpolating Control Based on Saturated Controllers

In this section, in order to fully utilize the capability of actuators and to extend the domain of attraction, an interpolation between several saturated controllers will be proposed. As in the previous section, we consider the case when $w(k) = 0$, $\forall k \geq 0$.

Fig. 5.12 State and input
trajectories of the closed loop
system as functions of time
for Example 5.3 for our
approach (*solid*), and for
[102] (*dashed*)

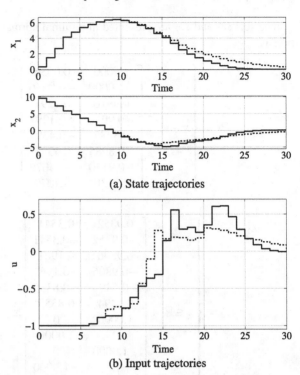

(a) State trajectories

(b) Input trajectories

For simplicity, only the single input–single output system case is considered here,
although extension to the multi-input multi-output system case is straightforward.

From Lemma 2.1 in Sect. 2.4.1, recall that for a given stabilizing controller
$u(k) = \text{sat}(Kx(k))$, there exists an auxiliary stabilizing controller $u(k) = Hx(k)$
such that the saturation function can be expressed as, $\forall x$ such that $Hx \in U$,

$$\text{sat}(Kx(k)) = \beta(k)Kx(k) + (1 - \beta(k))Hx(k) \tag{5.32}$$

where $0 \leq \beta(k) \leq 1$. The instrumental vector $H \in \mathbb{R}^n$ can be computed by Theorem 2.3. Based on Procedure 2.5 in Sect. 2.4.1, an associated robustly polyhedral
set Ω_s^H can be computed, that is invariant for the system,

$$x(k + 1) = A(k)x(k) + B(k)\,\text{sat}(Kx(k)) \tag{5.33}$$

It is assumed that a set of robustly asymptotically stabilizing controllers $u(k) =$
$\text{sat}(K_i x(k))$, $\forall i = 1, 2, \ldots, s$ is available as well as a set of auxiliary vectors $H_i \in$
$\mathbb{R}^n$ with $i = 2, 3, \ldots, s$ such that the maximal invariant set $\Omega_1 \subseteq X$

$$\Omega_1 = \left\{ x \in \mathbb{R}^n : F_o^{(1)} x \leq g_o^{(1)} \right\} \tag{5.34}$$

for the linear controller $u = K_1 x$, and the maximal invariant set $\Omega_s^{H_i} \subseteq X$

$$\Omega_s^{H_i} = \left\{ x \in \mathbb{R}^n : F_o^{(i)} x \leq g_o^{(i)} \right\} \tag{5.35}$$

Fig. 5.13 Interpolating
coefficient, Lyapunov
function and $\alpha(k)$ realization
as functions of time for
Example 5.3

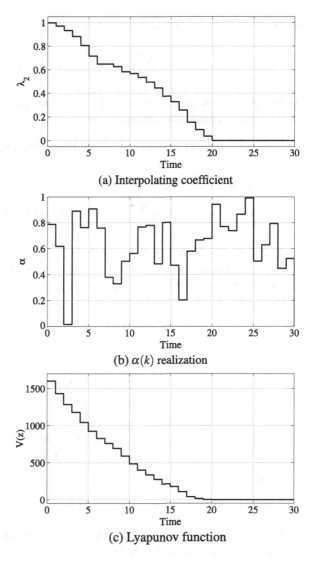

(a) Interpolating coefficient

(b) $\alpha(k)$ realization

(c) Lyapunov function

for $i = 2, 3, \ldots, s$ is non-empty. Denote Ω_s as the convex hull of the sets $\Omega_s^{H_i}$. By
the convexity of X, it follows that $\Omega_s \subseteq X$, since

$$\Omega_1 \subseteq X \quad \text{and} \quad \Omega_s^{H_i} \subseteq X, \quad \forall i = 2, 3 \ldots, s.$$

Any state $x(k) \in \Omega_s$ can be decomposed as,

$$x(k) = \lambda_1(k)\widehat{x}_1(k) + \sum_{i=2}^{s} \lambda_i(k)\widehat{x}_i(k) \tag{5.36}$$

where $\widehat{x}_1(k) \in \Omega_1, \widehat{x}_i(k) \in \Omega_s^{H_i}, \forall i = 2, 3, \ldots, s$ and

$$\sum_{i=1}^{s} \lambda_i(k) = 1, \quad \lambda_i(k) \geq 0$$

As in the previous section we remark the non-uniqueness of the decomposition (5.36). Consider the following control law,

$$u(k) = \lambda_1(k) K_1 \widehat{x}_1(k) + \sum_{i=2}^{s} \lambda_i(k) \operatorname{sat}(K_i \widehat{x}_i(k)) \tag{5.37}$$

Using Lemma 2.1, one obtains,

$$u(k) = \lambda_1(k) K_1 \widehat{x}_1(k) + \sum_{i=2}^{s} \lambda_i(k) \big(\beta_i(k) K_i + (1 - \beta_i(k)) H_i\big) \widehat{x}_i(k) \tag{5.38}$$

where $0 \leq \beta_i(k) \leq 1, \forall i = 2, 3, \ldots, s$.

With the same notation as in the previous section, let $r_i = \lambda_i \widehat{x}_i$. Since $\widehat{x}_1 \in \Omega_1$ and $\widehat{x}_i \in \Omega_s^{H_i}$, $i = 2, 3, \ldots, s$, it follows that $r_1 \in \lambda_1 \Omega_1$ and $r_i \in \lambda_i \Omega_s^{H_i}$, $i = 2, 3, \ldots, s$ or,

$$F_o^{(i)} x_i \leq \lambda_i g_o^{(i)}, \quad \forall i = 1, 2, \ldots, s \tag{5.39}$$

Using (5.36), (5.38), one gets,

$$\begin{cases} x = r_1 + \displaystyle\sum_{i=2}^{s} r_i, \\[4mm] u = u_1 + \displaystyle\sum_{i=2}^{s} u_i \end{cases} \tag{5.40}$$

where $u_1 = K_1 r_1$ and

$$u_i = (\beta_i K_i + (1 - \beta_i) H_i) r_i, \quad i = 2, 3, \ldots, s \tag{5.41}$$

The first high gain controller plays the role of a performance controller, while the remaining low gain controllers is used to enlarge the domain of attraction. It holds that,

$$x(k+1) = A(k)x(k) + B(k)u(k) = A(k) \sum_{i=1}^{s} r_i(k) + B(k) \sum_{i=1}^{s} u_i(k)$$

$$= \sum_{i=1}^{s} r_i(k+1)$$

where

$$r_1(k+1) = A(k)r_1(k) + B(k)u_1(k) = (A(k) + B(k)K_1)r_1(k)$$

and

$$r_i(k+1) = A(k)r_i(k) + B(k)u_i(k)$$
$$= \{A(k) + B(k)(\beta_i(k)K_i + (1 - \beta_i(k))H_i)\}r_i(k)$$

or

$$r_i(k+1) = A_{ci}(k)r_i(k) \tag{5.42}$$

with $A_{ci}(k) = A(k) + B(k)(\beta_i(k)K_i + (1 - \beta_i(k))H_i)$, $i = 2, 3, \ldots, s$.

For a given set of state and control weighting matrices $Q_i \succeq 0$ and $R_i \succ 0$, $i = 2, 3, \ldots, s$, consider the following set of quadratic functions,

$$V_i(r_i) = r_i^T P_i r_i, \quad i = 2, 3, \ldots, s \tag{5.43}$$

where matrix $P_i \in \mathbb{R}^{n \times n}$, $P_i \succ 0$ is chosen to satisfy,

$$V_i(r_i(k+1)) - V_i(r_i(k)) \leq -r_i(k)^T Q_i r_i(k) - u_i(k)^T R_i u_i(k) \tag{5.44}$$

Define $Y_i = \beta_i K_i + (1 - \beta_i)H_i$. From (5.41), (5.42), (5.43), one can rewrite inequality (5.44) as,

$$A_{ci}^T P_i A_{ci} - P_i \preceq -Q_i - Y_i^T R_i Y_i$$

By using the Schur complements, the previous condition can be transformed into,

$$\begin{bmatrix} P_i - Q_i - Y_i^T R_i Y_i & A_{ci}^T P_i \\ P_i A_{ci} & P_i \end{bmatrix} \succeq 0$$

or

$$\begin{bmatrix} P_i & A_{ci}^T P_i \\ P_i A_{ci} & P_i \end{bmatrix} - \begin{bmatrix} Q_i + Y_i^T R_i Y_i & 0 \\ 0 & 0 \end{bmatrix} \succeq 0$$

Denote $\sqrt{Q_i}$ and $\sqrt{R_i}$ as the Cholesky factor of the matrices Q_i and R_i, which satisfy $\sqrt{Q_i}^T \sqrt{Q_i} = Q_i$ and $\sqrt{R_i}^T \sqrt{R_i} = R_i$. The previous condition can be rewritten as,

$$\begin{bmatrix} P_i & A_{ci}^T P_i \\ P_i A_{ci} & P_i \end{bmatrix} - \begin{bmatrix} \sqrt{Q_i}^T & Y_i^T \sqrt{R_i}^T \\ 0 & 0 \end{bmatrix} \begin{bmatrix} \sqrt{Q_i} & 0 \\ \sqrt{R_i} Y_i & 0 \end{bmatrix} \succeq 0$$

or by using the Schur complements, one obtains,

$$\begin{bmatrix} P_i & A_{ci}^T P_i & \sqrt{Q_i}^T & Y_i^T \sqrt{R_i}^T \\ P_i A_{ci} & P_i & 0 & 0 \\ \sqrt{Q_i} & 0 & I & 0 \\ \sqrt{R_i} Y_i & 0 & 0 & I \end{bmatrix} \succeq 0 \tag{5.45}$$

Clearly, one should verify inequality (5.45) at the vertices of A_{ci} and Y_i. Hence the set of LMI conditions to be checked is the following,

$$\begin{cases} \begin{bmatrix} P_i & (A_j + B_j K_i)^T P_i & \sqrt{Q_i}^T & K_i^T \sqrt{R_i}^T \\ P_i(A_j + B_j K_i) & P_i & 0 & 0 \\ \sqrt{Q_i} & 0 & I & 0 \\ \sqrt{R_i} K_i & 0 & 0 & I \end{bmatrix} \succeq 0 \\ \begin{bmatrix} P_i & (A_j + B_j H_i)^T P_i & \sqrt{Q_i}^T & H_i^T \sqrt{R_i}^T \\ P_i(A_j + B_j H_i) & P_i & 0 & 0 \\ \sqrt{Q_i} & 0 & I & 0 \\ \sqrt{R_i} H_i & 0 & 0 & I \end{bmatrix} \succeq 0 \end{cases} \quad \forall j = 1, 2, \ldots, q$$

$$(5.46)$$

Condition (5.46) is linear with respect to the matrix P_i. One way to calculate P_i is to solve the following LMI problem,

$$\min_{P_i} \{ \text{trace}(P_i) \} \tag{5.47}$$

subject to constraint (5.46).

Once the matrices P_i with $i = 2, 3, \ldots, s$ are computed, they can be used in practice for real-time control based on the resolution of a low complexity optimization problem. At each time instant, for a given current state x, minimize on-line the following quadratic cost function subject to linear constraints,

$$\min_{r_i, \lambda_i} \left\{ \sum_{i=2}^{s} r_i^T P_i r_i + \sum_{i=2}^{s} \lambda_i^2 \right\} \tag{5.48}$$

subject to

$$\begin{cases} F_o^{(i)} r_i \leq \lambda_i g_o^{(i)}, \ \forall i = 1, 2, \ldots, s, \\ \sum_{i=1}^{s} r_i = x, \\ \sum_{i=1}^{s} \lambda_i = 1, \\ \lambda_i \geq 0, \forall i = 1, 2, \ldots, s \end{cases}$$

Theorem 5.4 *The control law (5.36), (5.37), (5.48) guarantees recursive feasibility and robustly asymptotic stability of the closed loop system for all initial states $x(0) \in \Omega_s$.*

Proof The proof is omitted here, since it is the same as the one of theorem 4.11. $\square$

Algorithm 5.3 Interpolating control via quadratic programming

---

1. Measure the current state $x(k)$.
2. Solve the QP problem (5.48).
3. Implement as input the control action $u = \lambda_1 K_1 \widehat{x}_1 + \sum_{i=2}^{s} \lambda_i \, \text{sat}(K_i \widehat{x}_i)$.
4. Wait for the next time instant $k := k + 1$.
5. Go to step 1 and repeat.

Fig. 5.14 Feasible invariant sets and state trajectories of the closed loop system for Example 5.4

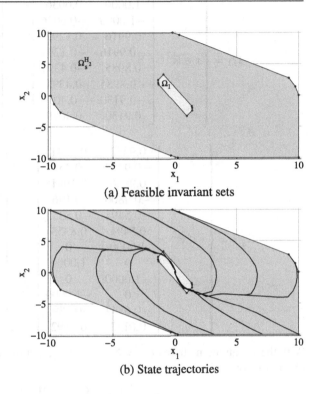

(a) Feasible invariant sets

(b) State trajectories

Example 5.4 We recall the uncertain, time-varying linear discrete-time system,

$$x(k+1) = \big(\alpha(k)A_1 + \big(1 - \alpha(k)\big)A_2\big)x(k) + \big(\alpha(k)B_1 + \big(1 - \alpha(k)\big)B_2\big)u(k) \quad (5.49)$$

in Example 5.1 with the same state and control constraints. Two matrix gains in the interpolation scheme are chosen as,

$$\begin{cases} K_1 = [-1.8112 \quad -0.8092], \\ K_2 = [-0.0936 \quad -0.1424] \end{cases} \quad (5.50)$$

Using Theorem 2.3, the matrix H_2 is computed as,

$$H_2 = [-0.0786 \quad -0.1010] \quad (5.51)$$

The robustly invariant sets Ω_1 and $\Omega_s^{H_2}$ are respectively, constructed for the controllers $u = K_1 x$ and $u = \text{sat}(K_2 x)$, see Fig. 5.14(a). Figure 5.14(b) shows state trajectories, obtained by solving the QP problem (5.48), for different initial conditions and realizations of $\alpha(k)$.

The sets Ω_1 and $\Omega_s^{H_2}$ are presented in minimal normalized half-space representation as,

$$\Omega_1 = \left\{ x \in \mathbb{R}^2 : \begin{bmatrix} 1.0000 & 0.0036 \\ -1.0000 & -0.0036 \\ 0.9916 & 0.1297 \\ -0.9916 & -0.1297 \\ 0.8985 & -0.4391 \\ -0.8985 & 0.4391 \\ -0.9130 & -0.4079 \\ 0.9130 & 0.4079 \end{bmatrix} x \le \begin{bmatrix} 1.3699 \\ 1.3699 \\ 1.1001 \\ 1.1001 \\ 2.3202 \\ 2.3202 \\ 0.5041 \\ 0.5041 \end{bmatrix} \right\}$$

$$\Omega_s^{H_2} = \left\{ x \in \mathbb{R}^2 : \begin{bmatrix} 0.9352 & 0.3541 \\ -0.9352 & -0.3541 \\ 0.9806 & 0.1961 \\ -0.9806 & -0.1961 \\ -0.5494 & -0.8355 \\ 0.5494 & 0.8355 \\ 1.0000 & 0 \\ 0 & 1.0000 \\ -1.0000 & 0 \\ 0 & -1.0000 \\ -0.6142 & -0.7892 \\ 0.6142 & 0.7892 \end{bmatrix} x \le \begin{bmatrix} 9.5779 \\ 9.5779 \\ 9.8058 \\ 9.8058 \\ 8.2385 \\ 8.2385 \\ 10.0000 \\ 10.0000 \\ 10.0000 \\ 10.0000 \\ 7.8137 \\ 7.8137 \end{bmatrix} \right\}$$

With the weighting matrices $Q_2 = I$, $R_2 = 10^{-4}$, and by solving the LMI problem (4.84), one obtains,

$$P_2 = \begin{bmatrix} 17.5066 & 7.9919 \\ 7.9919 & 16.7524 \end{bmatrix}$$

Using Algorithm 5.3, for the initial condition $x(0) = [-4.1194 \ 9.9800]^T$, Fig. 5.15 shows the state and input trajectories as functions of time for our approach (solid) and for the approach in [74] (dashed). Figure 5.16 presents the interpolating coefficient $\lambda_2(k)$, the objective function i.e. the Lyapunov function and the realization of $\alpha(k)$.

5.5 Interpolation via Quadratic Programming for Uncertain Systems with Disturbances

Note that all the developments in Sects. 5.3 and 5.4 avoided handling of additive disturbances due to the impossibility of dealing with the robustly asymptotic stabil-

Fig. 5.15 State and input trajectories as functions of time for our approach (*solid*) and for [74] (*dashed*) for Example 5.4

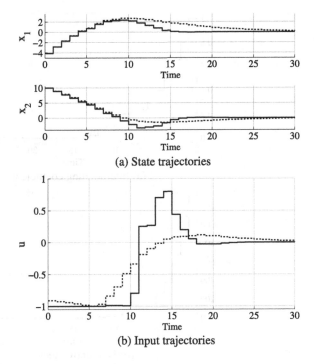

(a) State trajectories

(b) Input trajectories

ity of the origin as an equilibrium point. In this section, an interpolation for system (5.1) with constraints (5.3) using quadratic programming will be proposed to cope with the additive disturbance problem.

Clearly, when the disturbance is persistent, it is impossible to guarantee the convergence $x(k) \to 0$ as $k \to +\infty$. In other words, it is impossible to achieve asymptotic stability of the closed loop system to the origin. The best that can be hoped for is that the controller steers any initial state to some target set around the origin. Therefore an input-to-state stability (ISS) framework proposed in [59, 82, 91] will be used for characterizing this target region.

5.5.1 Input to State Stability

The input to state stability framework provides a natural way to formulate questions of stability with respect to disturbances [120]. This framework attempts to capture the notion of *bounded disturbance input—bounded state*.

Definition 5.1 ($\mathcal{K}$-function) A real valued scalar function $\phi : \mathbb{R}_{\geq 0} \to \mathbb{R}_{\geq 0}$ is of class $\mathcal{K}$ if it is continuous, strictly increasing and $\phi(0) = 0$.

Definition 5.2 ($\mathcal{K}_\infty$-function) A function $\phi : \mathbb{R}_{\geq 0} \to \mathbb{R}_{\geq 0}$ is of class $\mathcal{K}_\infty$ if it is a $\mathcal{K}$-function and $\phi(s) \to +\infty$ as $s \to +\infty$.

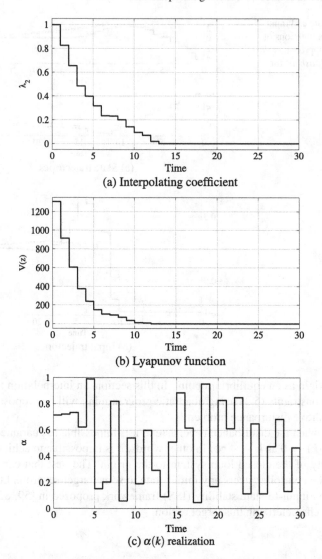

Fig. 5.16 Interpolating coefficient, Lyapunov function and $\alpha(k)$ realization as functions of time for Example 5.4

Definition 5.3 ($\mathscr{K}\mathscr{L}$-function) A function $\beta : \mathbb{R}_{\geq 0} \times \mathbb{R}_{\geq 0} \to \mathbb{R}_{\geq 0}$ is of class $\mathscr{K}\mathscr{L}$ if for each fixed $k \geq 0$, it follows that $\beta(\cdot, k)$ is a $\mathscr{K}$ function and for each fixed $s \geq 0$, it follows that $\beta(s, \cdot)$ is decreasing and $\beta(s, k) \to 0$ as $k \to 0$.

The ISS framework for autonomous uncertain/time-varying linear discrete-time systems, as studied by Jiang and Wang in [59], is briefly reviewed next.

Consider system (5.1) with a feedback controller $u(k) = Kx(k)$ and the corresponding closed loop matrix $A_c(k) = A(k) + B(k)K$,

$$x(k + 1) = A_c(k)x(k) + Dw(k) \qquad (5.52)$$

Definition 5.4 (ISS stability) The dynamical system (5.52) is ISS with respect to disturbance $w(k)$ if there exist a $\mathscr{K}\mathscr{L}$-function β and a $\mathscr{K}$-function ϕ such that for all initial states $x(0)$ and for all admissible disturbances $w(k)$, the evolution $x(k)$ of system (5.52) satisfies,

$$\|x(k)\| \leq \beta\big(\|x(0)\|, k\big) + \phi\Big(\sup_{0 \leq i \leq k-1} \|w(i)\|\Big) \qquad (5.53)$$

The function $\phi(\cdot)$ is usually called an ISS gain of system (5.52) [59, 91].

Definition 5.5 (ISS Lyapunov function) A function $V : \mathbb{R}^n \to \mathbb{R}_{\geq 0}$ is an ISS Lyapunov function for system (5.52) is there exist $\mathscr{K}_\infty$-functions $\gamma_1, \gamma_2, \gamma_3$ and a $\mathscr{K}$-function θ such that,

$$\begin{cases} \gamma_1\big(\|x\|\big) \leq V(x) \leq \gamma_2\big(\|x\|\big), \\ V\big(x(k+1)\big) - V\big(x(k)\big) \leq -\gamma_3\big(\|x(k)\|\big) + \theta\big(\|w(k)\|\big) \end{cases} \qquad (5.54)$$

Theorem 5.5 [59, 82, 91] *System* (5.52) *is input-to-state stable if it admits an ISS Lyapunov function.*

Remark 5.2 Note that the ISS notion is related to the existence of states x such that,

$$\gamma_3\big(\|x\|\big) \geq \theta\big(\|w\|\big)$$

for all $w \in W$. This implies that there exists a scalar $d \geq 0$ such that,

$$\gamma_3(d) = \max_{w \in W} \theta\big(\|w\|\big)$$

or $d = \gamma_3^{-1}(\max_{w(k) \in W} \theta(\|w(k)\|))$. Here γ_3^{-1} denotes the inverse operator of γ_3. It follows that for any $\|x(k)\| > d$, one has,

$$V\big(x(k+1)\big) - V\big(x(k)\big) \leq -\gamma_3\big(\|x(k)\|\big) + \theta\big(\|w(k)\|\big) \leq -\gamma_3(d) + \theta\big(\|w(k)\|\big) < 0$$

Thus the trajectory $x(k)$ of system (5.52) will eventually enter the region $R_x = \{x \in \mathbb{R}^n : \|x(k)\| \leq d\}$. Once inside, the trajectory will never leave this region, due to the monotonicity condition imposed on $V(x(k))$ outside the region R_x.

5.5.2 Cost Function Determination

The main contribution presented in the following starts from the assumption that a set of unconstrained robustly asymptotically stabilizing feedback controllers $u(k) =$

$K_i x(k)$, $i = 1, 2, \ldots, s$, is available such that for each i the matrix $A_{ci}(k)$ is asymptotically stable, where $A_{ci}(k) = A(k) + B(k)K_i$.

Using Procedure 2.2, the maximal robustly invariant set $\Omega_i \subseteq X$ is found as,

$$\Omega_i = \left\{ x \in \mathbb{R}^n : F_o^{(i)} x \leq g_o^{(i)} \right\} \tag{5.55}$$

for the corresponding controller $u(k) = K_i x(k)$, $i = 1, 2, \ldots, s$. With a slight abuse of notation, denote Ω as the convex hull of Ω_i, $i = 1, 2, \ldots, s$. It follows that $\Omega \subseteq X$, since X is convex and $\Omega_i \subseteq X$, $i = 1, 2, \ldots, s$.

Any state $x(k) \in \Omega$ can be decomposed as,

$$x(k) = \lambda_1(k)\widehat{x}_1(k) + \sum_{i=2}^{s} \lambda_i(k)\widehat{x}_i(k) \tag{5.56}$$

with $\widehat{x}_i(k) \in \Omega_i$ and $\sum_{i=1}^{s} \lambda_i(k) = 1$, $\lambda_i(k) \geq 0$.

One of the first remark is that according to the cardinal number s and the disposition of the sets Ω_i, the decomposition (5.56) is not unique.

Define $r_i(k) = \lambda_i(k)\widehat{x}_i(k)$, $\forall i = 1, 2, \ldots, s$. Equation (5.56) can be rewritten as,

$$x(k) = r_1(k) + \sum_{i=2}^{s} r_i(k)$$

or, equivalently

$$r_1(k) = x(k) - \sum_{i=2}^{s} r_i(k) \tag{5.57}$$

Since $\widehat{x}_i \in \Omega_i$, it follows that $r_i \in \lambda_i \Omega_i$, or in other words,

$$F_o^{(i)} r_i \leq \lambda_i g_o^{(i)}, \quad i = 1, 2, \ldots, s \tag{5.58}$$

Consider the following control law,

$$u(k) = \lambda_1(k)K_1\widehat{x}_1(k) + \sum_{i=2}^{s} \lambda_i(k)K_i\widehat{x}_i(k) = K_1 r_1(k) + \sum_{i=2}^{s} K_i r_i(k) \tag{5.59}$$

where $K_i\widehat{x}_i(k)$ is the control law in Ω_i.

Substituting (5.57) into (5.59), one gets,

$$u(k) = K_1 x(k) + \sum_{i=2}^{s} (K_i - K_1) r_i(k) \tag{5.60}$$

Using (5.60), one has,

$$x(k+1) = A(k)x(k) + B(k)u(k) + Dw(k)$$

$$= A(k)x(k) + B(k)K_1 x(k) + B(k) \sum_{i=2}^{s} (K_i - K_1) r_i(k) + Dw(k)$$

or, equivalently

$$x(k+1) = A_{c1}(k)x(k) + B(k) \sum_{i=2}^{s} (K_i - K_1) r_i(k) + Dw(k) \qquad (5.61)$$

where $A_{c1}(k) = A(k) + B(k)K_1$.

Define $r_i(k+1)$ as follows, $i = 2, 3, \ldots, s$,

$$r_i(k+1) = A_{ci}(k) r_i(k) + Dw_i(k) \qquad (5.62)$$

with $A_{ci}(k) = A(k) + B(k)K_i$ and $w_i(k) = \lambda_i(k)w(k)$.

Define the vectors z and ω as,

$$z = \begin{bmatrix} x^T & r_2^T & \ldots & r_s^T \end{bmatrix}^T, \qquad \omega = \begin{bmatrix} w^T & w_2^T & \ldots & w_s^T \end{bmatrix}^T \qquad (5.63)$$

Writing (5.61), (5.62) in a compact matrix form, one obtains,

$$z(k+1) = \Phi(k)z(k) + \Gamma\omega(k) \qquad (5.64)$$

where

$$\Phi(k) = \begin{bmatrix} A_{c1}(k) & B(k)(K_2 - K_1) & \ldots & B(k)(K_s - K_1) \\ 0 & A_{c2}(k) & \ldots & 0 \\ \vdots & \vdots & \ddots & \vdots \\ 0 & 0 & \ldots & A_{cs}(k) \end{bmatrix}, \qquad \Gamma = \begin{bmatrix} D & 0 & \ldots & 0 \\ 0 & D & \ldots & 0 \\ \vdots & \vdots & \ddots & \vdots \\ 0 & 0 & \ldots & D \end{bmatrix}$$

Using (5.2), it follows that $\Phi(k)$ can be expressed as a convex combination of Φ_j, i.e.

$$\begin{cases} \Phi(k) = \displaystyle\sum_{j=1}^{q} \alpha_j(k)\Phi_j, \\[2mm] \displaystyle\sum_{j=1}^{q} \alpha_j(k) = 1, \ \alpha_j(k) \geq 0 \end{cases} \qquad (5.65)$$

where

$$\Phi_j = \begin{bmatrix} (A_j + B_j K_1) & B_j(K_2 - K_1) & \ldots & B_j(K_s - K_1) \\ 0 & (A_j + B_j K_2) & \ldots & 0 \\ \vdots & \vdots & \ddots & \vdots \\ 0 & 0 & \ldots & (A_j + B_j K_s) \end{bmatrix}$$

Consider the following quadratic function,

$$V(z) = z^T P z \qquad (5.66)$$

where matrix $\mathscr{P} \succ 0$ is chosen to satisfy,

$$V\big(z(k+1)\big) - V\big(z(k)\big) \le -x(k)^T Q x(k) - u(k)^T R u(k) + \theta \omega(k)^T \omega(k) \quad (5.67)$$

where $Q \in \mathbb{R}^{n \times n}$, $R \in \mathbb{R}^{m \times m}$, $Q \succeq 0$ and $R \succ 0$ are the state and input weighting matrices, $\theta \ge 0$.

Using (5.64), the left hand side of (5.67) can be written as,

$$
\begin{aligned}
&V\big(z(k+1)\big) - V\big(z(k)\big) \\
&= (\Phi z + \Gamma \omega)^T P (\Phi z + \Gamma \omega) - z^T P z \\
&= \begin{bmatrix} z^T & \omega^T \end{bmatrix} \begin{bmatrix} \Phi^T \\ \Gamma^T \end{bmatrix} P \begin{bmatrix} \Phi & \Gamma \end{bmatrix} \begin{bmatrix} z \\ \omega \end{bmatrix} - \begin{bmatrix} z^T & \omega^T \end{bmatrix} \begin{bmatrix} P & 0 \\ 0 & 0 \end{bmatrix} \begin{bmatrix} z \\ \omega \end{bmatrix} \quad (5.68)
\end{aligned}
$$

And using (5.60), (5.63) the right hand side of (5.67) becomes,

$$
\begin{aligned}
&-x(k)^T Q x(k) - u(k)^T R u(k) + \theta \omega(k)^T \omega(k) \\
&= z(k)^T (-Q_1 - R_1) z(k) + \theta \omega(k)^T \omega(k) \\
&= \begin{bmatrix} z^T & \omega^T \end{bmatrix} \begin{bmatrix} -Q_1 - R_1 & 0 \\ 0 & \theta I \end{bmatrix} \begin{bmatrix} z \\ \omega \end{bmatrix} \quad (5.69)
\end{aligned}
$$

where

$$
Q_1 = \begin{bmatrix} I \\ 0 \\ \vdots \\ 0 \end{bmatrix} Q \begin{bmatrix} I & 0 & \cdots & 0 \end{bmatrix},
$$

$$
R_1 = \begin{bmatrix} K_1^T \\ (K_2 - K_1)^T \\ \vdots \\ (K_s - K_1)^T \end{bmatrix} R \begin{bmatrix} K_1 & (K_2 - K_1) & \cdots & (K_s - K_1) \end{bmatrix}
$$

From (5.67), (5.68), (5.69), one gets,

$$
\begin{bmatrix} \Phi^T \\ \Gamma^T \end{bmatrix} P \begin{bmatrix} \Phi & \Gamma \end{bmatrix} - \begin{bmatrix} P & 0 \\ 0 & 0 \end{bmatrix} \preceq \begin{bmatrix} -Q_1 - R_1 & 0 \\ 0 & \theta I \end{bmatrix}
$$

or equivalently,

$$
\begin{bmatrix} P - Q_1 - R_1 & 0 \\ 0 & \theta I \end{bmatrix} - \begin{bmatrix} \Phi^T \\ \Gamma^T \end{bmatrix} P \begin{bmatrix} \Phi & \Gamma \end{bmatrix} \succeq 0 \qquad (5.70)
$$

Using the Schur complements, equation (5.70) can be brought to,

$$\begin{bmatrix} P - Q_1 - R_1 & 0 & \Phi^T P \\ 0 & \theta I & \Gamma^T P \\ P\Phi & P\Gamma & P \end{bmatrix} \succeq 0 \qquad (5.71)$$

It is clear from (5.70) that problem (5.71) is feasible if the matrix $\Phi(k)$ is asymptotically stable, or in other words, all matrices $A_{ci}(k)$ are asymptotically stable.

The left hand side of (5.71) is linear with respect to $\alpha_j(k)$. Hence one should verify (5.71) at the vertices of $\alpha_j(k)$, i.e. when $\alpha_j(k) = 0$ or $\alpha_j(k) = 1$. Therefore the set of LMI conditions to be checked is as follows,

$$\begin{bmatrix} P - Q_1 - R_1 & 0 & \Phi_j^T P \\ 0 & \theta I & \Gamma^T P \\ P\Phi_j & P\Gamma & P \end{bmatrix} \succeq 0, \quad \forall j = 1, 2, \ldots, q \qquad (5.72)$$

Structurally, problem (5.72) is linear with respect to the matrix P and to the scalar θ. It is well known [82] that having a smaller θ is a desirable property in the sense of the ISS gain. The smallest value of θ can be computed by solving the following LMI optimization problem,

$$\min_{P,\theta} \{\theta\} \qquad (5.73)$$

subject to (5.72).

5.5.3 Interpolating Control via Quadratic Programming

Once the matrix P is computed as the solution of the problem (5.73), it can be used in practice for real time control based on the resolution of a low complexity optimization problem with respect to structure and complexity. The resulting control law can be seen as a predictive control type of construction, if the function (5.66) is interpreted as an upper bound for a receding horizon cost function.

Define the vector z_1 and the matrix P_1 as follows,

$$z_1 = \begin{bmatrix} x^T & r_2^T & \ldots & r_s^T & \lambda_2 & \lambda_3 & \ldots & \lambda_s \end{bmatrix}^T,$$

$$P_1 = \begin{bmatrix} P & 0 \\ 0 & I \end{bmatrix}$$

Consider the following quadratic function

$$J(z_1) = z_1^T P_1 z_1 \qquad (5.74)$$

At each time instant, for a given current state x, minimize on-line the following quadratic cost function,

$$V_1(z_1) = \min_{z_1}\{J(z_1)\} \tag{5.75}$$

subject to linear constraints

$$\begin{cases} F_o^{(i)}r_i \le \lambda_i g_o^{(i)}, \quad \forall i = 1, 2, \dots, s, \\[2mm] \displaystyle\sum_{i=1}^{s} r_i = x, \\[2mm] \lambda_i \ge 0, \quad \forall i = 1, 2, \dots, s, \\[2mm] \displaystyle\sum_{i=1}^{s} \lambda_i = 1 \end{cases}$$

and implement as input the control action $u = K_1 x + \sum_{i=2}^{s}(K_i - K_1)r_i$.

Theorem 5.6 *The control law (5.56), (5.60), (5.75) guarantees recursive feasibility and the closed loop system is ISS for all initial states $x(0) \in \Omega$.*

Proof Theorem 5.6 stands on two important claims, namely recursive feasibility and input-to-state stability. These can be treated sequentially.

Recursive feasibility: It has to be proved that $F_u u(k) \le g_u$ and $x(k+1) \in \Omega$ for all $x(k) \in \Omega$. Using (5.56), (5.59), one gets,

$$x(k) = \sum_{i=1}^{s} \lambda_i(k)\widehat{x}_i(k),$$

$$u(k) = \sum_{i=1}^{s} \lambda_i(k)K_i\widehat{x}_i(k)$$

It just holds that

$$F_u u(k) = F_u \sum_{i=1}^{s} \lambda_i(k)K_i\widehat{x}_i(k) = \sum_{i=1}^{s} \lambda_i(k)F_u K_i\widehat{x}_i(k)$$

$$\le \sum_{i=1}^{s} \lambda_i(k)g_u = g_u$$

and

$$x(k+1) = A(k)x(k) + B(k)u(k) + Dw(k)$$

$$= \sum_{i=1}^{s} \lambda_i(k)\{(A(k) + B(k)K_i)\widehat{x}_i(k) + Dw(k)\}$$

Since $(A(k) + B(k)K_i)\widehat{x}_i(k) + Dw(k) \in \Omega_i \subseteq \Omega$, it follows that $x(k+1) \in \Omega$.

ISS stability: From the feasibility proof, it is clear that if $r_i^*(k)$ and $\lambda_i^*(k)$, $i = 1, 2, \ldots, s$ is the solution of (5.75) at time k, then

$$r_i(k+1) = A_{ci}(k)r_i^*(k) + Dw_i(k)$$

and $\lambda_i(k+1) = \lambda_i^*(k)$ is a feasible solution of (5.75) at time $k+1$. Using (5.67), one obtains,

$$J\big(z_1(k+1)\big) - V_1\big(z_1(k)\big) \leq -x(k)^T Qx(k) - u(k)^T Ru(k) + \theta\omega(k)^T \omega(k) \quad (5.76)$$

By solving the QP problem (5.75) at time $k+1$, one gets,

$$V_1\big(z_1(k+1)\big) \leq J\big(z_1(k+1)\big)$$

It follows that,

$$V_1\big(z_1(k+1)\big) - V_1\big(z_1(k)\big) \leq J\big(z_1(k+1)\big) - V_1\big(z_1(k)\big)$$

$$\leq -x(k)^T Qx(k) - u(k)^T Ru(k) + \theta\omega(k)^T \omega(k)$$
$$(5.77)$$

where the last step of (5.77) follows from (5.76). Therefore $V_1(z_1)$ is an ISS Lyapunov function of the system (5.64). It follows that the closed loop system with the interpolating controller is ISS. □

Remark 5.3 Matrix P can be chosen as follows,

$$P = \begin{bmatrix} P_{11} & 0 \\ 0 & P_{ss} \end{bmatrix} \quad (5.78)$$

where $P_{11} \in \mathbb{R}^{n \times n}$, $P_{ss} \in \mathbb{R}^{(s-1)n \times (s-1)n}$. In this case, the cost function (5.75) can be written by,

$$J(z_1) = x^T P_{11}x + r_e^T P_{ss}r_e + \sum_{i=2}^{s} \lambda_i^2$$

where $r_e = [r_2^T \ r_3^T \ \ldots \ r_s^T]^T$. Hence for any $x \in \Omega_1$, the QP problem (5.75) has the trivial solution as,

$$\begin{cases} r_i = 0, \\ \lambda_i = 0 \end{cases} \quad \forall i = 2, 3, \ldots, s$$

and thus $r_1 = x$ and $\lambda_1 = 0$. Therefore, the interpolating controller becomes the optimal unconstrained controller $u = K_1 x$. It follows that the minimal robust positively invariant set R_∞ of the system,

$$x(k+1) = \big(A(k) + B(k)K_1\big)x(k) + Dw(k)$$

is an attractor of the closed loop system with the interpolating controller. In the other words, all trajectories will converge to the set R_∞.

Algorithm 5.4 Interpolating control via quadratic programming

1. Measure the current state $x(k)$.
2. Solve the QP problem (5.75).
3. Implement as input the control action $u = K_1 x + \sum_{i=2}^{r}(K_i - K_1)r_i$.
4. Wait for the next time instant $k := k + 1$.
5. Go to step 1 and repeat.

Example 5.5 Consider the following uncertain linear discrete-time system

$$x(k+1) = A(k)x(k) + B(k)u(k) + w(k) \qquad (5.79)$$

where

$$\begin{cases} A(k) = \alpha(k)A_1 + (1 - \alpha(k))A_2, \\ B(k) = \alpha(k)B_1 + (1 - \alpha(k))B_2 \end{cases}$$

and

$$A_1 = \begin{bmatrix} 1 & 0.1 \\ 0 & 1 \end{bmatrix}, \quad B_1 = \begin{bmatrix} 0 \\ 1 \end{bmatrix}, \quad A_2 = \begin{bmatrix} 1 & 0.2 \\ 0 & 1 \end{bmatrix}, \quad B_2 = \begin{bmatrix} 0 \\ 2 \end{bmatrix}$$

At each sampling time $\alpha(k) \in [0, 1]$ is an uniformly distributed pseudorandom number. The constraints are,

$$-10 \le x_1 \le 10, \qquad -10 \le x_2 \le 10, \qquad -1 \le u \le 1,$$

$$-0.1 \le w_1 \le 0.1, \qquad -0.1 \le w_2 \le 0.1$$

Three feedback controllers are chosen as,

$$\begin{cases} K_1 = [-1.8112 \quad -0.8092], \\ K_2 = [-0.0878 \quad -0.1176], \\ K_3 = [-0.0979 \quad -0.0499] \end{cases} \qquad (5.80)$$

With the weighting matrices $Q = I$, $R = 1$, and by solving the LMI problem (5.73) with P in the form (5.78), one obtains $\theta = 41150$ and With the weighting matrices $Q = I$, $R = 1$, and by solving the LMI problem (5.73) with P in the form (5.78), one obtains $\theta = 41150$ and

$$P_{11} = \begin{bmatrix} 76.2384 & 11.4260 \\ 11.4260 & 3.6285 \end{bmatrix}, \quad P_{33} = \begin{bmatrix} 2468.4 & 1622.9 & -144.2 & 105.2 \\ 1622.9 & 4164.3 & -81.6 & -160.6 \\ -144.2 & -81.6 & 865.7 & 278.4 \\ 105.2 & -160.6 & 278.4 & 967.8 \end{bmatrix}$$

Figure 5.17(a) shows the maximal robustly invariant sets Ω_1, Ω_2 and Ω_3, associated with the feedback gains K_1, K_2 and K_3, respectively. Figure 5.17(b) presents state trajectories for different initial conditions and realizations of $\alpha(k)$ and $w(k)$.

Fig. 5.17 Feasible invariant
sets and state trajectories of
the closed loop system for
Example 5.5

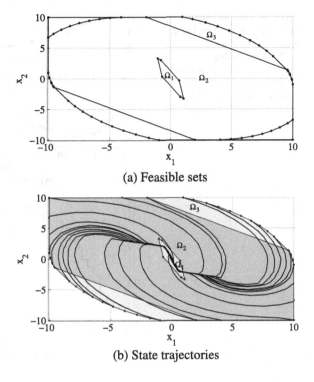

(a) Feasible sets

(b) State trajectories

For the initial condition $x_0 = [9.6145\ 1.1772]^T$ that is feasible for the controller $u(k) = K_3 x(k)$, Fig. 5.18 shows the state and input trajectories for the interpolating controller (solid) and for the controller $u(k) = K_3 x(k)$ (dashed). From Fig. 5.18, it is clear that the performance of the closed loop system with the interpolating controller is better than the performance of the closed loop system with the controller $u(k) = K_3 x(k)$.

Figure 5.19(a) presents the ISS Lyapunov function as a function of time. The non-decreasing phenomena of the ISS Lyapunov function, when the state is near to the origin is shown in Fig. 5.19(b).

Figure 5.20(a) shows the realization of $\alpha(k)$ and $w(k) = [w_1(k)\ w_2(k)]^T$ as functions of time. The interpolating coefficients $\lambda_2^*(k)$ and $\lambda_3^*(k)$ are depicted in Fig. 5.20(b).

5.6 Convex Hull of Invariant Ellipsoids for Uncertain, Time-Varying Systems

In this section, ellipsoids will be used as a class of sets for interpolation. It will be shown that the convex hull of a set of invariant ellipsoids is invariant. The ultimate goal is to design a method for constructing a continuous constrained feedback law based on interpolation for a given set of *saturated* control laws. In the absence of

Fig. 5.18 State and input
trajectories of the closed loop
system as functions of time
for the interpolating
controller (*solid*), and for the
controller $u(k) = K_3 x(k)$
(*dashed*) for Example 5.5

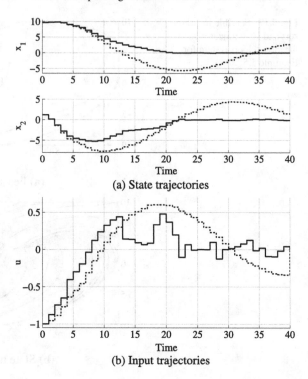

(a) State trajectories

(b) Input trajectories

disturbances, the system considered is of the form,

$$x(k+1) = A(k)x(k) + B(k)u(k) \qquad (5.81)$$

It is assumed that a set of asymptotically stabilizing saturated controllers $u =$
$\text{sat}(K_i x)$, $i = 1, 2, \ldots, s$ is available such that the corresponding robustly invariant
ellipsoids $E(P_i) \subset X$,

$$E(P_i) = \left\{ x \in \mathbb{R}^n : x^T P_i^{-1} x \le 1 \right\} \qquad (5.82)$$

are non-empty, $i = 1, 2, \ldots, s$. Recall that for all $x(k) \in E(P_i)$, it follows that
$\text{sat}(K_i x) \in U$ and

$$x(k+1) = A(k)x(k) + B(k)\,\text{sat}\big(K_i x(k)\big) \in E(P_i)$$

Denote Ω_E as the convex hull of $E(P_i)$. It follows that $\Omega_E \subseteq X$, since X is convex
and $E(P_i) \subseteq X$, $i = 1, 2, \ldots, s$.

Any state $x(k) \in \Omega_E$ can be decomposed as follows,

$$x(k) = \sum_{i=1}^{s} \lambda_i(k)\widehat{x}_i(k) \qquad (5.83)$$

Fig. 5.19 ISS Lyapunov function and its non-increasing phenomena as functions of time for Example 5.5

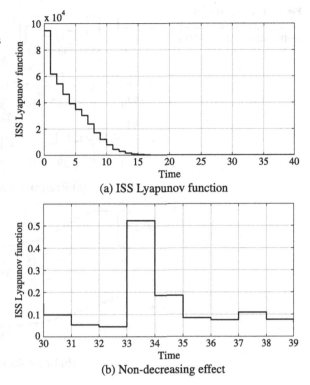

(a) ISS Lyapunov function

(b) Non-decreasing effect

with $\widehat{x}_i(k) \in E(P_i)$ and $\lambda_i(k)$ are the interpolating coefficients, that satisfy,

$$\sum_{i=1}^{s} \lambda_i = 1, \quad \lambda_i \geq 0$$

Consider the following control law,

$$u(k) = \sum_{i=1}^{s} \lambda_i(k) \, \mathrm{sat}\left(K_i \widehat{x}_i(k)\right) \tag{5.84}$$

where $u_i(k) = \mathrm{sat}(K_i \widehat{x}_i(k))$ is the saturated control law in $E(P_i)$.

Theorem 5.7 *The control law* (5.83), (5.84) *guarantees recursive feasibility for all* $x(0) \in \Omega_E$.

Proof The proof of this theorem is the same as the proof of Theorem 4.13 and is omitted here.   □

   As in the previous Sections, the first feedback gain in the sequence is used for satisfying performance specifications near the origin, while the remaining gains are

Fig. 5.20 Realization of $\alpha(k)$, $w(k)$ and interpolating coefficients $\lambda_2^*(k)$ and $\lambda_3^*(k)$ for Example 5.5

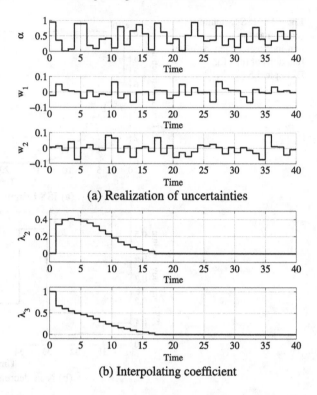

(a) Realization of uncertainties

(b) Interpolating coefficient

used to enlarge the domain of attraction. For a given current state x, consider the following optimization problem,

$$\lambda_i^* = \min_{\widehat{x}_i, \lambda_i} \left\{ \sum_{i=2}^{s} \lambda_i \right\} \tag{5.85}$$

subject to,

$$\begin{cases} \widehat{x}_i^T P_i^{-1} \widehat{x}_i \leq 1, \quad \forall i = 1, 2, \ldots, s, \\[2mm] \displaystyle\sum_{i=1}^{s} \lambda_i \widehat{x}_i = x, \\[2mm] \displaystyle\sum_{i=1}^{s} \lambda_i = 1, \\[2mm] \lambda_i \geq 0, \quad \forall i = 1, 2, \ldots, s \end{cases}$$

Theorem 5.8 *The control law* (5.83), (5.84), (5.85) *guarantees robustly asymptotic stability for all initial states* $x(0) \in \Omega_E$.

Proof Consider the following positive function,

$$V(x) = \sum_{i=2}^{s} \lambda_i^*(x) \tag{5.86}$$

for all $x \in \Omega_E \setminus E(P_1)$. $V(x)$ is a Lyapunov function candidate.

At time k by solving the optimization problem (5.85) and by applying (5.83), (5.84), one has, for any $x(k) \in \Omega_E \setminus E(P_1)$,

$$\begin{cases} x(k) = \sum_{i=1}^{s} \lambda_i^*(k)\widehat{x}_i^*(k), \\ u(k) = \sum_{i=1}^{s} \lambda_i^*(k) \operatorname{sat}\left(K_i \widehat{x}_i^*(k)\right) \end{cases}$$

It follows that,

$$x(k+1) = A(k)x(k) + B(k)u(k)$$

$$= A(k) \sum_{i=1}^{s} \lambda_i^*(k)\widehat{x}_i^*(k) + B(k) \sum_{i=1}^{s} \lambda_i^*(k) \operatorname{sat}\left(K_i \widehat{x}_i^*(k)\right)$$

$$= \sum_{i=1}^{s} \lambda_i^*(k)\widehat{x}_i(k+1)$$

where

$$\widehat{x}_i(k+1) = A(k)\widehat{x}_i^*(k) + B(k) \operatorname{sat}\left(K_i \widehat{x}_i^*(k)\right) \in E(P_i), \quad i = 1, 2, \dots, s$$

Hence $\lambda_i^*(k)$, $i = 1, 2, \dots, s$ is a feasible solution of (5.85) at time $k + 1$.

By solving the optimization problem (5.85) at time $k + 1$, the optimal solution is obtained,

$$x(k+1) = \sum_{i=1}^{s} \lambda_i^*(k+1)\widehat{x}_i^*(k+1)$$

where $\widehat{x}_i^*(k+1) \in E(P_i)$. It follows that,

$$\sum_{i=2}^{s} \lambda_i^*(k+1) \leq \sum_{i=2}^{s} \lambda_i^*(k)$$

and $V(x)$ is a non-increasing function.

The contractive invariant property of ellipsoids $E(P_i)$, $i = 1, 2, \dots, s$ assures that there is no initial condition $x(0) \in \Omega_E \setminus E(P_1)$ such that $\sum_{i=2}^{s} \lambda_i^*(k+1) = \sum_{i=2}^{s} \lambda_i^*(k)$ for all $k \geq 0$. It follows that $V(x) = \sum_{i=2}^{s} \lambda_i^*(x)$ is a Lyapunov function for all $x \in \Omega_E \setminus E(P_1)$.

Algorithm 5.5 Interpolating control—Convex hull of ellipsoids

---

1. Measure the current state $x(k)$.
2. Solve the LMI problem (5.87).
3. Apply as input the control law (5.84).
4. Wait for the next instant $k := k + 1$.
5. Go to step 1 and repeat.

The proof is complete by noting that inside $E(P_1)$, $\lambda_1^* = 1$, the robustly stabilizing controller $u = \mathrm{sat}(K_1 x)$ is contractive and thus the interpolating controller assures robustly asymptotic stability for all $x \in \Omega_E$.   □

Define $r_i = \lambda_i \widehat{x}_i$. Since $\widehat{x}_i \in E(P_i)$, it follows that $r_i^T P_i^{-1} r_i \leq \lambda_i^2$ or by using the Schur complements,

$$\begin{bmatrix} \lambda_i & r_i^T \\ r_i & \lambda_i P_i \end{bmatrix} \succeq 0$$

Hence, the non-linear optimization problem (5.85) can be brought to the following LMI optimization problem,

$$\lambda_i^* = \min_{r_i, \lambda_i} \left\{ \sum_{i=2}^{s} \lambda_i \right\} \qquad (5.87)$$

subject to

$$\begin{cases} \begin{bmatrix} \lambda_i & r_i^T \\ r_i & \lambda_i P_i \end{bmatrix} \succeq 0, & \forall i = 1, 2, \ldots, s, \\[2mm] \sum_{i=1}^{s} r_i = x, \\[2mm] \sum_{i=1}^{s} \lambda_i = 1, \\[2mm] \lambda_i \geq 0, & \forall i = 1, 2, \ldots, s \end{cases}$$

For $x \in E(P_1)$, it is clear that the optimization problem (5.85) has the trivial solution

$$\lambda_i^* = 0, \quad i = 2, 3, \ldots, s.$$

And hence $\lambda_1^* = 1$ and $\widehat{x}_1 = x$. It follows that the interpolating controller becomes the saturated controller $u = \mathrm{sat}(K_1 x)$ inside $E(P_1)$.

Fig. 5.21 Feasible invariant sets and state trajectories of the closed loop system for Example 5.6

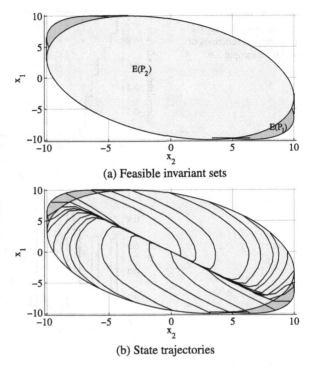

(a) Feasible invariant sets

(b) State trajectories

Fig. 5.22 State and input trajectories of the closed loop system as functions of time for Example 5.6

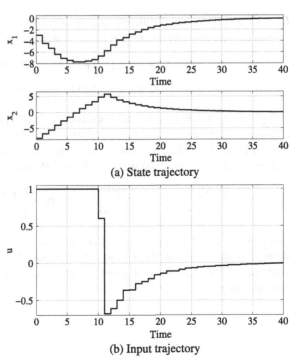

(a) State trajectory

(b) Input trajectory

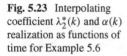

Fig. 5.23 Interpolating
coefficient $\lambda_2^*(k)$ and $\alpha(k)$
realization as functions of
time for Example 5.6

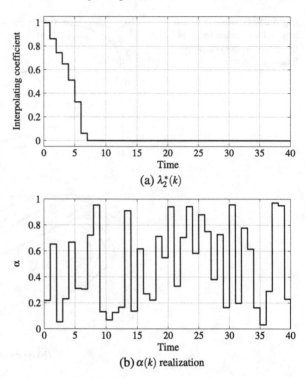

(a) $\lambda_2^*(k)$

(b) $\alpha(k)$ realization

Example 5.6 Consider the uncertain system in Example 5.1 with the same state and control constraints. Two matrix gains are chosen as,

$$\begin{cases} K_1 = [-0.6451 \quad -0.7740], \\ K_2 = [-0.2416 \quad -0.7824] \end{cases} \tag{5.88}$$

Using Theorem 2.3, two invariant ellipsoids $E(P_1)$ and $E(P_2)$ are respectively, constructed for the saturated controllers $u = \text{sat}(K_1 x)$ and $u = \text{sat}(K_2 x)$, see Fig. 5.21(a). Figure 5.21(b) shows state trajectories for different initial conditions and realizations of $\alpha(k)$. The matrices P_1 and P_2 are,

$$P_1 = \begin{bmatrix} 100.0000 & -64.4190 \\ -64.4190 & 100.00000 \end{bmatrix}, \quad P_2 = \begin{bmatrix} 100.0000 & -32.2659 \\ -32.2659 & 100.0000 \end{bmatrix}$$

Using Algorithm 5.5, for the initial condition $x(0) = [-2.96 \;\; -8.08]^T$, Fig. 5.22 shows the state and the input trajectories. Figure 5.23 presents the interpolating coefficient $\lambda_2^*(k)$ and the realization of $\alpha(k)$. As expected, $\lambda_2^*(k)$ is positive and non-increasing.

Chapter 6
Interpolating Control—Output Feedback Case

6.1 Problem Formulation

Consider the problem of regulating to the origin the following uncertain and/or time-varying linear discrete-time system, described by the input-output relationship,

$$y(k+1) + E_1 y(k) + E_2 y(k-1) + \cdots + E_s y(k-s+1)$$
$$= N_1 u(k) + N_2 u(k-1) + \cdots + N_r u(k-r+1) + w(k) \qquad (6.1)$$

where $y(k) \in \mathbb{R}^p$, $u(k) \in \mathbb{R}^m$ and $w(k) \in \mathbb{R}^p$ are respectively, the output, the input and the disturbance vector. The matrices $E_i \in \mathbb{R}^{p \times p}$, $i = 1, 2, \ldots, s$ and $N_j \in \mathbb{R}^{p \times m}$, $j = 1, 2, \ldots, r$.

For simplicity, it is assumed that $s = r$. The matrices E_i and N_i, $i = 1, 2, \ldots, s$ satisfy,

$$\begin{bmatrix} E_1 & E_2 & \cdots & E_s \\ N_1 & N_2 & \cdots & N_s \end{bmatrix} = \sum_{j=1}^{q} \alpha_j(k) \begin{bmatrix} E_1^{(j)} & E_2^{(j)} & \cdots & E_s^{(j)} \\ N_1^{(j)} & N_2^{(j)} & \cdots & N_s^{(j)} \end{bmatrix} \qquad (6.2)$$

where $\alpha_j(k) \geq 0$ and $\sum_{j=1}^{q} \alpha_j(k) = 1$ and

$$\begin{bmatrix} E_1^{(j)} & E_2^{(j)} & \cdots & E_s^{(j)} \\ N_1^{(j)} & N_2^{(j)} & \cdots & N_s^{(j)} \end{bmatrix}, \quad j = 1, 2, \ldots, q$$

are the extreme realizations of the polytopic model (6.2).

The output, control and disturbance vectors are subject to the following bounded polytopic constraints,

$$\begin{cases} y(k) \in Y, & Y = \{ y \in \mathbb{R}^p : F_y y \leq g_y \}, \\ u(k) \in U, & U = \{ u \in \mathbb{R}^m : F_u u \leq g_u \}, \\ w(k) \in W, & W = \{ w \in \mathbb{R}^p : F_w w \leq g_w \}, \end{cases} \qquad (6.3)$$

where the matrices F_y, F_u, F_w and the vectors g_y, g_u, g_w are assumed to be constant with $g_y > 0$, $g_u > 0$ and $g_w > 0$.

H.-N. Nguyen, *Constrained Control of Uncertain, Time-Varying, Discrete-Time Systems*,   159
Lecture Notes in Control and Information Sciences 451,
DOI 10.1007/978-3-319-02827-9_6,
© Springer International Publishing Switzerland 2014

6.2 Output Feedback—Nominal Case

In this section, we consider the case when the matrices E_i and N_i for $i = 1, 2, \ldots, s$ are known and fixed. The case when E_i and N_i for $i = 1, 2, \ldots, s$ are uncertain and/or time-varying will be treated in the next section.

A state space representation will be constructed along the lines of [126]. All the steps of the construction are detailed such that the presentation are self contained. The state of the system is chosen as follows,

$$x(k) = \begin{bmatrix} x_1(k)^T & x_2(k)^T & \ldots & x_s(k)^T \end{bmatrix}^T \tag{6.4}$$

where

$$\begin{cases} x_1(k) = y(k) \\ x_2(k) = -E_s x_1(k-1) + N_s u(k-1) \\ x_3(k) = -E_{s-1} x_1(k-1) + x_2(k-1) + N_{s-1} u(k-1) \\ x_4(k) = -E_{s-2} x_1(k-1) + x_3(k-1) + N_{s-2} u(k-1) \\ \vdots \\ x_s(k) = -E_2 x_1(k-1) + x_{s-1}(k-1) + N_2 u(k-1) \end{cases} \tag{6.5}$$

The components of the state vector can be interpreted exclusively in terms of the input and output vectors as,

$$x_2(k) = -E_s y(k-1) + N_s u(k-1)$$

$$x_3(k) = -E_{s-1} y(k-1) - E_s y(k-2) + N_{s-1} u(k-1) + N_s u(k-2)$$

$$\vdots$$

$$x_s(k) = -E_2 y(k-1) - E_3 y(k-2) - \cdots - E_s y(k-s+1)$$
$$\qquad + N_2 u(k-1) + N_3 u(k-2) + \cdots + N_s u(k-s+1)$$

It holds that,

$$y(k+1) = -E_1 y(k) - E_2 y(k-1) - \cdots - E_s y(k-s+1)$$
$$\qquad + N_1 u(k) + N_2 u(k-1) + \cdots + N_s u(k-s+1) + w(k)$$

or, equivalently

$$x_1(k+1) = -E_1 x_1(k) + x_s(k) + N_1 u(k) + w(k)$$

The state space model is then defined in a compact form as follows,

$$\begin{cases} x(k+1) = Ax(k) + Bu(k) + Dw(k) \\ y(k) = Cx(k) \end{cases} \tag{6.6}$$

where

$$A = \begin{bmatrix} -E_1 & 0 & 0 & \cdots & 0 & I \\ -E_s & 0 & 0 & \cdots & 0 & 0 \\ -E_{s-1} & I & 0 & \cdots & 0 & 0 \\ -E_{s-2} & 0 & I & \cdots & 0 & 0 \\ \vdots & \vdots & \vdots & \ddots & \vdots & \vdots \\ -E_2 & 0 & 0 & \cdots & I & 0 \end{bmatrix}, \quad B = \begin{bmatrix} N_1 \\ N_s \\ N_{s-1} \\ N_{s-2} \\ \vdots \\ N_2 \end{bmatrix}, \quad D = \begin{bmatrix} I \\ 0 \\ 0 \\ 0 \\ \vdots \\ 0 \end{bmatrix}$$

and

$$C = \begin{bmatrix} I & 0 & 0 & 0 & \cdots & 0 \end{bmatrix}$$

Clearly, the realization (6.6) is *minimal* in the single-input single-output case. However, in the multi-input multi-output case, this realization might not be minimal, as shown in the following example.

Consider the following single-input multi-output linear discrete-time system,

$$y(k+1) - \begin{bmatrix} 2 & 0 \\ 0 & 2 \end{bmatrix} y(k) + \begin{bmatrix} 1 & 0 \\ 0 & 1 \end{bmatrix} y(k-1)$$

$$= \begin{bmatrix} 0.5 \\ 2 \end{bmatrix} u(k) + \begin{bmatrix} 0.5 \\ 1 \end{bmatrix} u(k-1) + w(k) \tag{6.7}$$

Using the construction (6.4), (6.5), the state space model is given as,

$$\begin{cases} x(k+1) = Ax(k) + Bu(k) + Dw(k) \\ y(k) = Cx(k) \end{cases}$$

where

$$A = \begin{bmatrix} 2 & 0 & 1 & 0 \\ 0 & 2 & 0 & 1 \\ -1 & 0 & 0 & 0 \\ 0 & -1 & 0 & 0 \end{bmatrix}, \quad B = \begin{bmatrix} 0.5 \\ 0.5 \\ 0.5 \\ -1.5 \end{bmatrix},$$

$$D = \begin{bmatrix} 1 \\ 1 \\ 0 \\ 0 \end{bmatrix}, \quad C = \begin{bmatrix} 1 & 0 & 0 & 0 \\ 0 & 1 & 0 & 0 \end{bmatrix}$$

This realization is not minimal, since it unnecessarily replicates the common poles of the denominator in the input-output description. There exists minimal state space realization like,

$$A = \begin{bmatrix} 0 & -1 \\ 1 & 2 \end{bmatrix}, \quad B = \begin{bmatrix} 0.5 \\ 0.5 \end{bmatrix}, \quad D = \begin{bmatrix} 0 \\ 1 \end{bmatrix}, \quad C = \begin{bmatrix} 0 & 1 \\ 1 & 0 \end{bmatrix}$$

Define

$$z(k) = \begin{bmatrix} y(k)^T & \cdots & y(k-s+1)^T & u(k-1)^T & \cdots & u(k-s+1)^T \end{bmatrix}^T \tag{6.8}$$

Using (6.5), the state $x(k)$ is expressed through $z(k)$ as,

$$x(k) = Tz(k) \tag{6.9}$$

where $T = [T_1 \ T_2]$ and

$$T_1 = \begin{bmatrix} I & 0 & 0 & \cdots & 0 \\ 0 & -E_s & 0 & \cdots & 0 \\ 0 & -E_{s-1} & -E_s & \cdots & 0 \\ \vdots & \vdots & \vdots & \ddots & \vdots \\ 0 & -E_2 & -E_3 & \cdots & -E_s \end{bmatrix}, \qquad T_2 = \begin{bmatrix} 0 & 0 & 0 & \cdots & 0 \\ N_s & 0 & 0 & \cdots & 0 \\ N_{s-1} & N_s & 0 & \cdots & 0 \\ \vdots & \vdots & \vdots & \ddots & \vdots \\ N_2 & N_3 & N_4 & \cdots & N_s \end{bmatrix}$$

Hence, it becomes obvious that at any time instant k, the state vector is available exclusively though measured input and output variables and their past values.

Using (6.3), (6.5), it follows that the state constraints are $x_i \in X_i$, where X_i are given as,

$$\begin{cases} X_1 = Y, \\ X_2 = E_s(-X_1) \oplus N_s U, \\ X_i = E_{s+2-i}(-X_1) \oplus X_{i-1} \oplus N_{s+2-i} U, \quad \forall i = 3, \ldots, s \end{cases} \tag{6.10}$$

Example 6.1 Consider the following discrete-time system,

$$y(k+1) - 2y(k) + y(k-1) = 0.5u(k) + 0.5u(k-1) + w(k) \tag{6.11}$$

The constraints are,

$$-5 \le y(k) \le 5, \qquad -5 \le u(k) \le 5, \qquad -0.1 \le w(k) \le 0.1$$

Using the construction (6.4), (6.5), the state space model is given as,

$$\begin{cases} x(k+1) = Ax(k) + Bu(k) + Dw(k) \\ y(k) = Cx(k) \end{cases}$$

where

$$A = \begin{bmatrix} 2 & 1 \\ -1 & 0 \end{bmatrix}, \qquad B = \begin{bmatrix} 0.5 \\ 0.5 \end{bmatrix}, \qquad E = \begin{bmatrix} 1 \\ 0 \end{bmatrix}, \qquad C = \begin{bmatrix} 1 & 0 \end{bmatrix}$$

$x(k)$ is available though the measured input, output and their past values as,

$$x(k) = \begin{bmatrix} 1 & 0 & 0 \\ 0 & -1 & 0.5 \end{bmatrix} \begin{bmatrix} y(k) \\ y(k-1) \\ u(k-1) \end{bmatrix}$$

Using (6.10), the constraints on the state are,

$$-5 \le x_1 \le 5, \qquad -7.5 \le x_2 \le 7.5$$

The local controller is chosen as an LQ controller with the following weighting matrices,

$$Q = C^T C = \begin{bmatrix} 1 & 0 \\ 0 & 0 \end{bmatrix}, \qquad R = 0.1$$

giving the state feedback gain,

$$K = \begin{bmatrix} -2.3548 & -1.3895 \end{bmatrix}$$

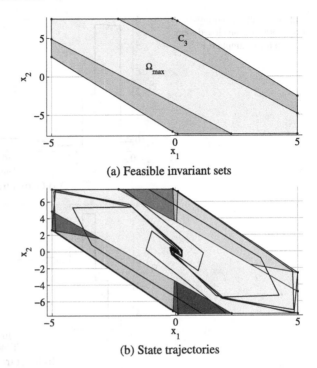

Fig. 6.1 Feasible invariant sets and state trajectories for Example 6.1

(a) Feasible invariant sets

(b) State trajectories

This example will use Algorithm 5.1 in Sect. 5.2, where vertex control is a global controller. Using Procedure 2.2 and Procedure 2.3, the sets Ω_{max} and C_N with $N = 3$ are found and shown in Fig. 6.1(a). Note that $C_3 = C_4$ is the maximal invariant set for system (6.11). Figure 6.1(b) presents state trajectories for different initial conditions and realizations of $w(k)$.

The set of vertices of C_N is given by the matrix $V(C_N)$ below, together with the control matrix U_v,

$$V(C_N) = \begin{bmatrix} -5 & -0.1 & 5 & 0.1 & -0.1 & -5 & 0.1 & 5 \\ 7.5 & 7.5 & -2.6 & 7.2 & -7.2 & 2.6 & -7.5 & -7.5 \end{bmatrix},$$

$$U_v = \begin{bmatrix} -5 & -5 & -5 & -4.9 & 5 & 5 & 5 & 4.9 \end{bmatrix}$$

Ω_{max} is presented in minimal normalized half-space representation as,

$$\Omega_{max} = \left\{ x \in \mathbb{R}^2 : \begin{bmatrix} 1.0000 & 0 \\ 0 & 1.0000 \\ -1.0000 & 0 \\ 0 & -1.0000 \\ -0.8612 & -0.5082 \\ 0.8612 & 0.5082 \end{bmatrix} x \le \begin{bmatrix} 5.0000 \\ 7.5000 \\ 5.0000 \\ 7.5000 \\ 1.8287 \\ 1.8287 \end{bmatrix} \right\}$$

For the initial condition $x(0) = [-0.1000 \quad 7.5000]^T$, Fig. 6.2 shows the output and input trajectories as functions of time.

Fig. 6.2 Output and input
trajectories of the closed loop
system for Example 6.1

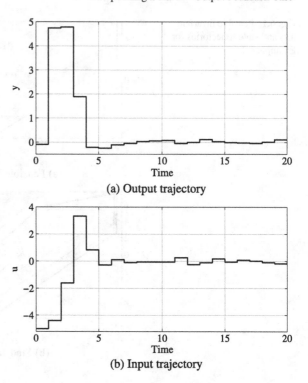

Fig. 6.2 Output and input
trajectories of the closed loop
system for Example 6.1

(a) Output trajectory

(b) Input trajectory

The interpolating coefficient and the realization of $w(k)$ as functions of time are depicted in Fig. 6.3. As expected, the interpolating coefficient, i.e. the Lyapunov function is positive and non-increasing.

As a comparison, we present a solution based on the well-known steady state Kalman filter. Figure 6.4 shows the output trajectories for the constrained output feedback approach (solid) and for the Kalman filter + constrained state feedback approach (dashed).

The *Kalman* function of Matlab 2011b was used for designing the Kalman filter. The process noise is a white noise with an uniform distribution and no measurement noise was considered. The disturbance w is a random number with an uniform distribution, $w_l \leq w \leq w_u$ where $w_l = -0.1$ and $w_u = 0.1$. The variance of w is given as,

$$C_w = \frac{(w_u - w_l + 1)^2 - 1}{12} = 0.0367$$

The estimator gain of the Kalman filter is obtained as,

$$L = [2 \quad -1]^T$$

The Kalman filter is used to estimate the state of and then this estimation is used to close the loop with the interpolating controller. In contrast to the output feedback approach, where the state is exact with respect to the measurement, in the Kalman

Fig. 6.3 Interpolating
coefficient and realization of
$w(k)$ for Example 6.1

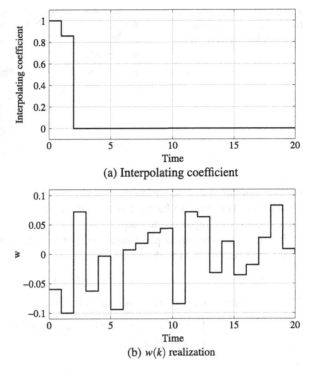

(a) Interpolating coefficient

(b) $w(k)$ realization

Fig. 6.4 Output trajectories
for our approach (*solid*) and
for the Kalman filter based
approach (*dashed*) for
Example 6.1

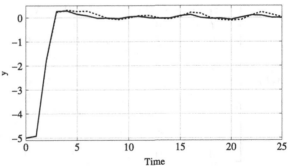

filter approach, an extra level of uncertainty is introduced, since the real state is unknown. Thus there is no guarantee that the constraints are satisfied in the transitory stage. This constraint violation effect is shown in Fig. 6.5.

6.3 Output Feedback—Robust Case

A weakness of the approach in Sect. 6.2 is that the state measurement is available if and only if the parameters of the system are known. For uncertain and/or time-varying system, that is not the case. In this section, we provide another method

Fig. 6.5 Constraint violation
for the Kalman filter based
approach for Example 6.1

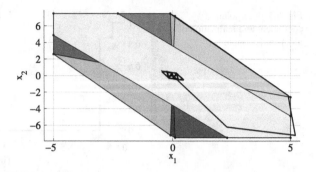

for constructing the state variables, that do not use the information of the system
parameters. The price to be paid is that the realization is in general *non-minimal*
even in the single-input single-output case.

Based on the measured plant input, output and their past measured values, the
state of the system (6.1) is chosen as,

$$x(k) = \begin{bmatrix} y(k)^T & \dots & y(k-s+1)^T & u(k-1)^T & \dots & u(k-s+1)^T \end{bmatrix}^T \quad (6.12)$$

The state space model is then defined as follows,

$$\begin{cases} x(k+1) = A(k)x(k) + B(k)u(k) + Dw(k) \\ y(k) = Cx(k) \end{cases} \quad (6.13)$$

where

$$A(k) = \begin{bmatrix} -E_1(k) & -E_2(k) & \dots & -E_s(k) & N_2(k) & \dots & N_{s-1}(k) & N_s(k) \\ I & 0 & \dots & 0 & 0 & \dots & 0 & 0 \\ 0 & I & \dots & 0 & 0 & \dots & 0 & 0 \\ \vdots & \vdots & \ddots & \vdots & \vdots & \ddots & \vdots & \vdots \\ 0 & 0 & \dots & I & 0 & \dots & 0 & 0 \\ 0 & 0 & \dots & 0 & 0 & \dots & 0 & 0 \\ 0 & 0 & \dots & O & I & \dots & 0 & 0 \\ \vdots & \vdots & \ddots & \vdots & \vdots & \ddots & \vdots & \vdots \\ 0 & 0 & \dots & O & 0 & \dots & I & 0 \end{bmatrix}$$

$$B(k) = \begin{bmatrix} N_1(k)^T & 0 & 0 & \dots & 0 & I & 0 & \dots & 0 \end{bmatrix}^T$$

$$D = \begin{bmatrix} I & 0 & 0 & \dots & 0 & 0 & 0 & \dots & 0 \end{bmatrix}^T$$

$$C = \begin{bmatrix} I & 0 & 0 & \dots & 0 & 0 & 0 & \dots & 0 \end{bmatrix}$$

Using (6.2), it follows that matrices $A(k)$ and $B(k)$ belong to a polytopic set,

$$(A, B) \in \text{Conv}\{(A_1, B_1), (A_2, B_2), \dots, (A_q, B_q)\} \quad (6.14)$$

where the vertices (A_i, B_i), $i = 1, 2, \dots, q$ are obtained from the vertices of (6.2).

Although the obtained representation is non-minimal, it has the merit that the
original output-feedback problem for the uncertain and/or time-varying plant has
been transformed into a state-feedback problem where the matrices A and B lie in

the polytope defined by (6.14) without any additional uncertainty. Clearly, any state-feedback control which is designed for the representation (6.13) in the form $u = Kx$ can be translated into a dynamic output-feedback controller.

Using (6.3), it follows that $x(k) \in X \subset \mathbb{R}^{s \times (p+m)}$, where the set X is given by,

$$X = \underbrace{Y \times Y \times \cdots \times Y}_{s \text{ times}} \times \underbrace{U \times U \times \cdots \times U}_{s \text{ times}}$$

Example 6.2 Consider the following transfer function,

$$P(s) = \frac{k_1 s + 1}{s(s + k_2)} \tag{6.15}$$

where $k_1 = 0.787, 0.1 \leq k_2 \leq 3$. Using a sampling time of 0.1 and Euler's first order approximation for the derivative, the following input-output relationship is obtained,

$$y(k+1) - (2 - 0.1k_2)y(k) + (1 - 0.1k_2)y(k-1)$$
$$= 0.1k_1 u(k) + (0.01 - 0.1k_2)u(k-1) + w(k) \tag{6.16}$$

The signal $w(k)$ is added to represent the process noise with $-0.01 \leq w \leq 0.01$. The constraints on output and input are,

$$-10 \leq y \leq 10, \qquad -5 \leq u \leq 5$$

The state $x(k)$ is constructed as follows,

$$x(k) = \begin{bmatrix} y(k) & y(k-1) & u(k-1) \end{bmatrix}^T$$

Hence, the state space model is given by,

$$\begin{cases} x(k+1) = A(k)x(k) + Bu(k) + Dw(k) \\ y(k) = Cx(k) \end{cases}$$

where

$$A(k) = \begin{bmatrix} (2 - 0.1k_2) & -(1 - 0.1k_2) & (0.01 - 0.1k_1) \\ 1 & 0 & 0 \\ 0 & 0 & 0 \end{bmatrix},$$

$$B = \begin{bmatrix} 0.1k_1 \\ 0 \\ 1 \end{bmatrix}, \qquad D = \begin{bmatrix} 1 \\ 0 \\ 0 \end{bmatrix} \quad \text{and} \quad C = \begin{bmatrix} 1 & 0 & 0 \end{bmatrix}$$

Using the polytopic uncertainty description, one obtains,

$$A(k) = \alpha(k)A_1 + (1 - \alpha(k))A_2$$

where

$$A_1 = \begin{bmatrix} 1.99 & -0.99 & -0.0687 \\ 1 & 0 & 0 \\ 0 & 0 & 0 \end{bmatrix}, \qquad A_2 = \begin{bmatrix} 1.7 & -0.7 & -0.0687 \\ 1 & 0 & 0 \\ 0 & 0 & 0 \end{bmatrix}$$

At each time instant $0 \leq \alpha(k) \leq 1$ and $-0.01 \leq w(k) \leq 0.01$ are uniformly distributed pseudo-random numbers. This example will use Algorithm 5.1 with a global saturated controller. For this purpose, two controllers have been designed

Fig. 6.6 Feasible invariant
sets for Example 6.2

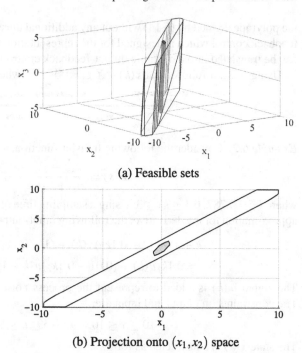

(a) Feasible sets

(b) Projection onto (x_1, x_2) space

Fig. 6.7 Output and input
trajectories of the closed loop
system for Example 6.2

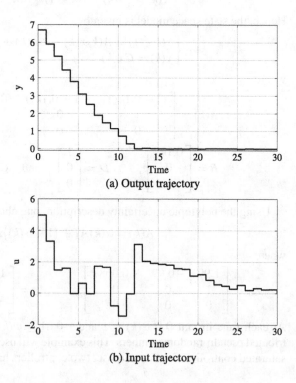

(a) Output trajectory

(b) Input trajectory

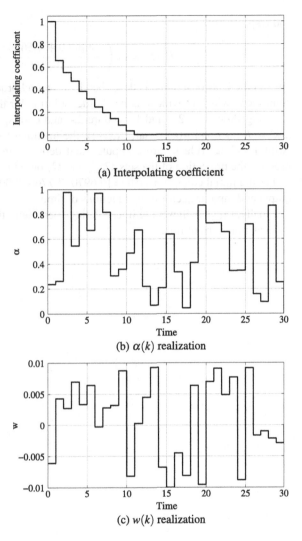

Fig. 6.8 Interpolating coefficient and realizations of $\alpha(k)$ and $w(k)$ for Example 6.2

(a) Interpolating coefficient

(b) $\alpha(k)$ realization

(c) $w(k)$ realization

- The local linear controller $u(k) = Kx(k)$ for the performance is chosen as,

$$K = [-22.7252 \quad 10.7369 \quad 0.8729]$$

- The global saturated controller $u(k) = \mathrm{sat}(K_s x(k))$ for the domain of attraction,

$$K_s = [-4.8069 \quad 4.5625 \quad 0.3365]$$

It is worth noticing that $u(k) = Kx(k)$ and $u(k) = \mathrm{sat}(K_s x(k))$ can be described in the output-feedback form as,

$$K(z) = \frac{-22.7894 + 10.7369z^{-1}}{1 - 0.8729z^{-1}}$$

and respectively

$$K_s(z) = \text{sat}\left(\frac{-4.8069 + 4.5625z^{-1}}{1 - 0.3365z^{-1}}\right)$$

Overall the control scheme is described by a second order plant and two first order controllers, which provide a reduced order solution for the stabilization problem.

Using Procedure 2.2 and Procedure 2.4 and corresponding to the control laws $u(k) = Kx(k)$ and $u(k) = \text{sat}(K_s x(k))$, the maximal robustly invariant sets Ω_{max} (white) and Ω_s (black) are computed and depicted in Fig. 6.6(a). Figure 6.6(b) presents the projection of the sets Ω_{max} and Ω_s onto the (x_1, x_2) state space.

For the initial condition $x(0) = [6.6970 \ 7.7760 \ 5.0000]^T$, Fig. 6.7 presents the output and input trajectories as functions of time.

Finally, Fig. 6.8 shows the interpolating coefficient, the realizations of $\alpha(k)$ and $w(k)$ as functions of time.

Part III
Applications

Chapter 7
High-Order Examples

7.1 Implicit Interpolating Control for High Order Systems

This example is taken from [11]. We consider the problem of regulating the yaw
and lateral dynamics in heavy vehicles, consisting of combinations of truck and
multiple towed units, see Fig. 7.1 and Fig. 7.2. In such heavy vehicle configurations,
undesired yaw rate and lateral acceleration amplifications, causing tail swings and
lateral instabilities of the towed units, can be observed at high speed [11].

It is well known [11, 44, 116] that this vehicle model can be described by a
linear parameter varying system, where the longitudinal velocity v_x is the parameter
variation. By choosing $v_x = 80$ km/h, the following system is obtained,

$$\begin{cases} x(k+1) = Ax(k) + Bu(k), \\ y(k) = Cx(k) \end{cases} \tag{7.1}$$

where

$$\begin{cases} x = [x_1 \quad x_2 \quad x_3 \quad x_4 \quad x_5 \quad x_6]^T, \\ y = [y_1 \quad y_2 \quad y_3 \quad y_4]^T \end{cases} \tag{7.2}$$

with $x_1 = v_{y1}$ is the lateral velocity of the truck, $x_2 = \dot{\psi}_1$ is the yaw rate of the
truck, $x_3 = \theta_1$ and $x_4 = \theta_2$ are the articulation angles of the dolly and the semitrailer,
$x_5 = \dot{\theta}_1$ and $x_6 = \dot{\theta}_2$ are, respectively the derivatives of x_3 and x_4, y_1 and y_2 are the
yaw rates of the dolly and semitrailer, $y_3 = x_3$ and $y_4 = x_4$, $u_1 = \delta_2$, $u_2 = \delta_3$ are the
lumped (relative) steering angles of the dolly and the semitrailer, respectively. The
matrices A, B, C are,

$$A = \begin{bmatrix} 0.8668 & -1.6664 & -0.0118 & 0.0002 & -0.0681 & 0.0007 \\ 0.0038 & 0.8604 & 0.0043 & -0.0001 & 0.0243 & -0.0002 \\ -0.0046 & 0.0018 & 0.8173 & 0.0032 & -1.0296 & 0.0100 \\ 0.0002 & 0.0507 & 0.0472 & 0.8440 & 1.0278 & -0.4982 \\ -0.0002 & 0.0000 & 0.0639 & 0.0001 & 0.9627 & 0.0004 \\ 0.0000 & 0.0019 & 0.0013 & 0.0646 & 0.0373 & 0.9821 \end{bmatrix},$$

H.-N. Nguyen, *Constrained Control of Uncertain, Time-Varying, Discrete-Time Systems*, 173
Lecture Notes in Control and Information Sciences 451,
DOI 10.1007/978-3-319-02827-9_7,
© Springer International Publishing Switzerland 2014

Fig. 7.1
Truck-dolly-semitrailer

Fig. 7.2
Truck-dolly-semitrailer:
schematic model. The *first
three blocks* represent the
wheels of the truck, the
second two blocks represent
the wheels of the dolly, and
the *last three blocks* represent
the wheels of the semitrailer

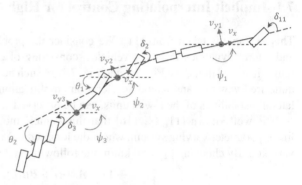

$$B = \begin{bmatrix} -0.0688 & 0.0246 & -1.0396 & 1.5260 & -0.0376 & 0.0552 \\ 0.0007 & -0.0002 & 0.0100 & -0.4982 & 0.0004 & -0.0179 \end{bmatrix}^{T},$$

$$C = \begin{bmatrix} 0 & 1 & 0 & 0 & 1 & 0 \\ 0 & 1 & 0 & 0 & 1 & 1 \\ 0 & 0 & 1 & 0 & 0 & 0 \\ 0 & 0 & 0 & 1 & 0 & 0 \end{bmatrix}$$

$$(7.3)$$

The input and output constraints are

$$-8 \leq y_1 \leq 8, \quad -8 \leq y_2 \leq 8, \quad -8 \leq y_3 \leq 8,$$
$$-8 \leq y_4 \leq 8, \quad -0.5 \leq u_1 \leq 0.5, \quad -0.5 \leq u_2 \leq 0.5 \tag{7.4}$$

Algorithm 5.1 in Sect. 5.2 will be used in this example, where the following
saturated controller $u = \text{sat}(K_s x)$,

$$K_s = \begin{bmatrix} 0.0017 & -0.0160 & 0.0067 & -0.0227 & -0.1199 & -0.0012 \\ -0.0010 & 0.0084 & 0.0070 & 0.0221 & 0.1102 & -0.0333 \end{bmatrix} \tag{7.5}$$

is used as a global controller.

Fig. 7.3 Illustration of the feasible invariant sets for the controller $u = Kx$ (*black*) and for the controller $u = \text{sat}(K_s x)$ (*white*)

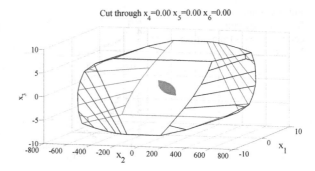

The local controller is chosen as a linear quadratic (LQ) controller with weighting matrices

$$Q = C^T \begin{bmatrix} 2 & 0 & 0 & 0 \\ 0 & 4 & 0 & 0 \\ 0 & 0 & 2 & 0 \\ 0 & 0 & 0 & 2 \end{bmatrix} \quad C = \begin{bmatrix} 0 & 0 & 0 & 0 & 0 & 0 \\ 0 & 6 & 0 & 0 & 6 & 4 \\ 0 & 0 & 2 & 0 & 0 & 0 \\ 0 & 0 & 0 & 2 & 0 & 0 \\ 0 & 6 & 0 & 0 & 6 & 4 \\ 0 & 4 & 0 & 0 & 4 & 4 \end{bmatrix}, \quad R = \begin{bmatrix} 0.01 & 0 \\ 0 & 0.01 \end{bmatrix}$$

(7.6)

giving the state feedback gain,

$$K = \begin{bmatrix} 0.0057 & 0.4661 & 0.8373 & 0.0194 & 0.3996 & 0.6023 \\ 0.0291 & 2.1493 & 2.6772 & 1.8025 & 4.5305 & 3.1107 \end{bmatrix} \quad (7.7)$$

Using Procedure 2.2 and Procedure 2.4 the sets Ω_{max} and Ω_s are respectively, computed for the control laws $u(k) = Kx(k)$ and $u(k) = \text{sat}(K_s x(k))$. The set Ω_s is found after 5 iterations in Procedure 2.4. Ω_{max} and Ω_s are illustrated in Fig. 7.3.

For the MPC setup, we take MPC, based on quadratic programming, where an LQ criterion is optimized, with the same weighting matrices as in (7.6). Hence the set Ω_{max} for the local unconstrained control is identical for the MPC solution and for the implicit interpolating controller. The prediction horizon for the MPC was chosen to be 40 to match the set Ω_s.

For the initial condition

$$x(0) = [-185.6830 \quad 9.1019 \quad 1.1120 \quad 8.0000 \quad -1.1019 \quad -4.9729]',$$

Fig. 7.4 and Fig. 7.5 show the output and input trajectories as functions of time for the interpolating controller (solid) and for the MPC controller (dashed).

Using the tic/toc function of Matlab 2011b, the computational burdens of interpolating controller and MPC were compared. The result is shown in Table 7.1.

Finally, Fig. 7.6 presents the interpolating coefficient as a function of time. As expected, this function is positive and decreasing.

Fig. 7.4 Output trajectories for the interpolating controller (*solid*) and for the MPC controller (*dashed*)

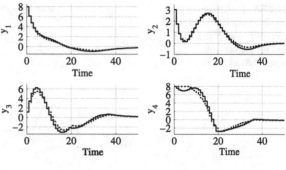

Fig. 7.5 Input trajectories for the interpolating controller (*solid*) and for the MPC controller (*dashed*)

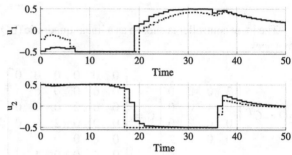

Fig. 7.6 Interpolating coefficient as a function of time

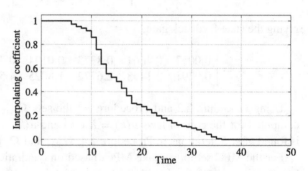

Table 7.1 Comparison between interpolating control and model predictive control

	Solver	Number of decision variables	Duration [ms] for the on-line computations during one sampling interval
Implicit Interpolating Controller	LP	7	0.9869
Implicit MPC	QP	80	28.9923

7.2 Explicit Interpolating Controller for High Order Systems

This example is taken from [122]. Consider the following discrete-time linear time-invariant system

$$x(k+1) = Ax(k) + Bu(k) \tag{7.8}$$

Fig. 7.7 Illustration of the feasible invariant sets for the controller $u = Kx$ (*black*) and for the controller $u = \text{sat}(K_sx)$ (*white*)

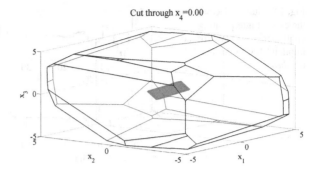

Cut through $x_4 = 0.00$

where

$$A = \begin{bmatrix} 0.4035 & 0.3704 & 0.2935 & -0.7258 \\ -0.2114 & 0.6405 & -0.6717 & -0.0420 \\ 0.8368 & 0.0175 & -0.2806 & 0.3808 \\ -0.0724 & 0.6001 & 0.5552 & 0.4919 \end{bmatrix},$$

$$B = \begin{bmatrix} 1.6124 & 0.4086 & -1.4512 & -0.6761 \end{bmatrix}^T$$

(7.9)

The constraints are

$$-5 \leq x_1 \leq 5, \quad -5 \leq x_2 \leq 5, \quad -5 \leq x_3 \leq 5,$$
$$-5 \leq x_4 \leq 5, \quad -0.2 \leq u \leq 0.2$$

(7.10)

For our explicit interpolating control solution, we choose local LQ control with the following weighting matrices

$$Q = \begin{bmatrix} 1 & 0 & 0 & 0 \\ 0 & 1 & 0 & 0 \\ 0 & 0 & 1 & 0 \\ 0 & 0 & 0 & 1 \end{bmatrix}, \quad R = 0.2$$

(7.11)

The calculated local feedback gain is then

$$K = \begin{bmatrix} -0.0047 & -0.1082 & -0.0496 & 0.4110 \end{bmatrix}$$

(7.12)

The following saturated controller $u = \text{sat}(K_sx)$,

$$K_s = \begin{bmatrix} 0.0104 & -0.0064 & -0.0100 & 0.0439 \end{bmatrix}$$

(7.13)

is chosen as a global controller.

A cut of the maximal invariant set Ω_{max} for the controller $u = Kx$ and the invariant set Ω_s for the saturated controller $u = \text{sat}(K_sx)$ are shown in Fig. 7.7. The set Ω_s is computed by using Procedure 2.4 after 10 iterations.

The Explicit Interpolating Controller is computed. It covers 945 cells and is not given here but will be sent to the reader on demand. Note that one does not need to

Fig. 7.8 State trajectories for the explicit interpolating controller (*solid*) and for the implicit MPC controller (*dashed*)

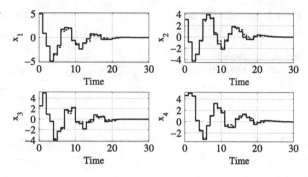

Fig. 7.9 Input trajectories for the explicit interpolating controller (*solid*) and for the implicit MPC controller (*dashed*)

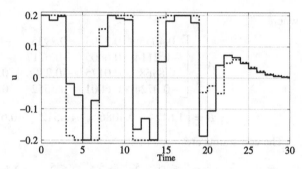

Fig. 7.10 Interpolating coefficient as a function of time

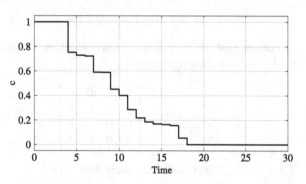

perform step 2 in Algorithm 4.2, since the saturated controller (7.13) is chosen as a global controller.

For the initial condition

$$x(0) = [5.0000 \quad 3.0625 \quad 2.5060 \quad 4.6109]^T,$$

Fig. 7.8 and Fig. 7.9 show the state and input trajectories for the explicit interpolating controller.

Figure 7.8 and Fig. 7.9 also show the state and input trajectories obtained by using implicit MPC with quadratic cost, with the weighting matrices as in (7.11) and with prediction horizon = 17.

As reported in [122], the explicit solution of the QP-MPC problem with the prediction horizon 17 and with the weighting matrices (7.11) could not be fully computed for this example due to the high complexity. The solution was terminated when 50000 polyhedral cells were already computed. In [122] using adaptive multiscale bases, an approximate explicit MPC solution was constructed, that consists of 3633 regions.

The interpolating coefficient as a function of time is depicted in Fig. 7.10.

Chapter 8
A Benchmark Problem: The Non-isothermal Continuous Stirred Tank Reactor

8.1 Continuous Stirred Tank Reactor Model

The case of a single non-isothermal continuous stirred tank reactor [69, 90, 111] is studied in this chapter. The reactor is the one presented in various works by Perez et al. [98, 99] in which the exothermic reaction $\mathscr{A} \to \mathscr{B}$ is assumed to take place. The heat of reaction is removed via the cooling jacket that surrounds the reactor. The jacket cooling water is assumed to be perfectly mixed and the mass of the metal walls is considered negligible, so that the thermal inertia of the metal is not considered. The reactor is also assumed to be perfectly mixed and heat losses are regarded as negligible, see Fig. 8.1.

The continuous linearized reactor system [90] is modeled as,

$$\dot{x} = A_c x + B_c u \tag{8.1}$$

where $x = [x_1 \ x_2]^T$, x_1 is the reactor concentration and x_2 is the reactor temperature, $u = [u_1 \ u_2]^T$, u_1 is the feed concentration and u_2 is the coolant flow. The matrices A_c and B_c are,

$$A_c = \begin{bmatrix} \left(-\frac{F}{V} - k_0(t)e^{-\frac{E}{RT_s}}\right) & \left(-\frac{E}{RT_s^2}k_0(t)e^{-\frac{E}{RT_s}}C_{As}\right) \\ \left(-\frac{\Delta H_{rxn}(t)k_0(t)e^{-\frac{E}{RT_s}}}{\rho C_p}\right) & \left(-\frac{F}{V} - \frac{UA}{V\rho C_p} - \Delta H_{rxn}(t)\frac{E}{\rho C_p RT_s^2}k_0(t)e^{-\frac{E}{RT_s}}C_{As}\right) \end{bmatrix},$$

$$B_c = \begin{bmatrix} \frac{F}{V} & 0 \\ 0 & -2.098 \times 10^5 \frac{T_s - 365}{V\rho C_p} \end{bmatrix}$$

$$\tag{8.2}$$

The operating parameters are shown in Table 8.1.

The linearized model at steady state $x_1 = 0.265$ kmol/m^3 and $x_2 = 394$ K and under the uncertain parameters k_0 and $-\Delta H_{rxn}$ will be considered. The following uncertain system [130] is obtained after discretizing system (8.1) with a sampling

H.-N. Nguyen, *Constrained Control of Uncertain, Time-Varying, Discrete-Time Systems,*   181
Lecture Notes in Control and Information Sciences 451,
DOI 10.1007/978-3-319-02827-9_8,
© Springer International Publishing Switzerland 2014

Fig. 8.1 Continuous stirred tank reactor

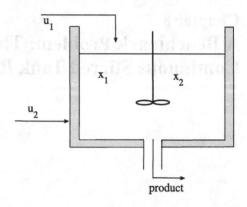

Table 8.1 The operating parameters of non-isothermal CSTR

Parameter	Value	Unit
F	1	m^3/min
V	1	m^3
ρ	10^6	g/m^3
C_p	1	cal/g·K
ΔH_{rxn}	10^7–10^8	cal/kmol
E/R	8330.1	K
k_o	10^9–10^{10}	min^{-1}
UA	5.34×10^6	cal/K·min

time of 0.15 min,

$$\begin{cases} x(k+1) = A(k)x(k) + Bu(k) \\ y(k) = Cx(k) \end{cases} \tag{8.3}$$

where

$$A(k) = \begin{bmatrix} 0.85 - 0.0986\beta_1(k) & -0.0014\beta_1(k) \\ 0.9864\beta_1(k)\beta_2(k) & 0.0487 + 0.01403\beta_1(k)\beta_2(k) \end{bmatrix},$$

$$B = \begin{bmatrix} 0.15 & 0 \\ 0 & -0.912 \end{bmatrix}, \qquad C = \begin{bmatrix} 1 & 0 \\ 0 & 1 \end{bmatrix}$$

and the parameter variation bounded by,

$$\begin{cases} 1 \le \beta_1(k) = \dfrac{k_0}{10^9} \le 10 \\ 1 \le \beta_2(k) = -\dfrac{\Delta H_{rxn}}{10^7} \le 10 \end{cases}$$

Matrix $A(k)$ can be expressed as,

$$A(k) = \alpha_1(k)A_1 + \alpha_2(k)A_2 + \alpha_3(k)A_3 + \alpha_4(k)A_4$$

Fig. 8.2 Feasible invariant sets

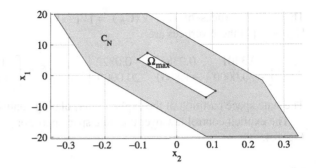

where $\sum_{i=1}^{4} \alpha_i(k) = 1$, $\alpha_i(k) \geq 0$ and

$$A_1 = \begin{bmatrix} 0.751 & -0.0014 \\ 0.986 & 0.063 \end{bmatrix}, \qquad A_2 = \begin{bmatrix} 0.751 & -0.0014 \\ 9.864 & 0.189 \end{bmatrix}$$

$$A_3 = \begin{bmatrix} -0.136 & -0.014 \\ 9.864 & 0.189 \end{bmatrix}, \qquad A_4 = \begin{bmatrix} -0.136 & -0.014 \\ 98.644 & 1.451 \end{bmatrix}$$

The input and state constraints on input are,

$$\begin{cases} -0.5 \leq x_1 \leq 0.5, & -20 \leq x_2 \leq 20, \\ -0.5 \leq u_1 \leq 0.5, & -1 \leq u_2 \leq 1 \end{cases} \tag{8.4}$$

8.2 Controller Design

The explicit interpolating controller in Sect. 5.2 will be used in this example. The local feedback controller $u(k) = Kx(k)$ is chosen as,

$$K = \begin{bmatrix} -2.8413 & 0.0366 \\ 34.4141 & 0.5195 \end{bmatrix} \tag{8.5}$$

Based on Procedure 2.2 and Procedure 2.3, the robustly maximal invariant set $\Omega_{\max}$ and the robustly controlled invariant set C_N with $N = 9$ are computed. Note that $C_9 = C_{10}$ is the maximal controlled invariant set for system (8.3) with constraints (8.4). The sets $\Omega_{\max}$ and C_N are depicted in Fig. 8.2.

The set $\Omega_{\max}$ given in half-space representation is,

$$\Omega_{\max} = \left\{ x \in \mathbb{R}^2 : \begin{bmatrix} -1.0000 & 0.0129 \\ 1.0000 & 0.0151 \\ 1.0000 & -0.0129 \\ -1.0000 & -0.0151 \end{bmatrix} x \leq \begin{bmatrix} 0.2501 \\ 0.1760 \\ 0.0291 \\ 0.1760 \\ 0.0291 \end{bmatrix} \right\}$$

The set of vertices of C_N, $V(C_N) = [V_1 \ -V_1]$, and the control matrix $U_v = [U_1 \ -U_1]$ at these vertices are,

$$V_1 = \begin{bmatrix} 0.3401 & 0.2385 & -0.0822 \\ -20.0000 & -1.8031 & 20.0000 \end{bmatrix}, \quad U_1 = \begin{bmatrix} -0.5000 & -0.5000 & 0.3534 \\ 1.0000 & 1.0000 & 1.0000 \end{bmatrix}$$

The state space partition of the explicit interpolating controller is shown in Fig. 8.3.

The explicit control law over the state space partition, see below, is illustrated in Fig. 8.4.

$$u(k) = \begin{cases}
\begin{bmatrix} -0.50 \\ 1.00 \end{bmatrix} & \text{if} & \begin{bmatrix} 1.00 & 0.01 \\ -1.00 & -0.02 \\ -1.00 & 0.04 \end{bmatrix} x(k) \leq \begin{bmatrix} 0.23 \\ -0.03 \\ -0.31 \end{bmatrix} \\[12pt]
\begin{bmatrix} -0.75 & 0.03 \\ 0.00 & 0.00 \end{bmatrix} x(k) + \begin{bmatrix} -0.27 \\ 1.00 \end{bmatrix} & \text{if} & \begin{bmatrix} 1.00 & -0.04 \\ 1.00 & 0.01 \\ -1.00 & -0.01 \end{bmatrix} x(k) \leq \begin{bmatrix} 0.31 \\ 0.21 \\ -0.07 \end{bmatrix} \\[12pt]
\begin{bmatrix} -6.05 & -0.01 \\ 0.00 & 0.00 \end{bmatrix} x(k) + \begin{bmatrix} 0.09 \\ 1 \end{bmatrix} & \text{if} & \begin{bmatrix} -1.00 & -0.02 \\ 1.00 & 0.01 \\ -1.00 & -0.00 \end{bmatrix} x(k) \leq \begin{bmatrix} -0.03 \\ 0.07 \\ 0.08 \end{bmatrix} \\[12pt]
\begin{bmatrix} -0.57 & -0.01 \\ 7.75 & 0.00 \end{bmatrix} x(k) + \begin{bmatrix} 0.54 \\ 1.63 \end{bmatrix} & \text{if} & \begin{bmatrix} 1.00 & 0.00 \\ -1.00 & -0.02 \\ 0.00 & 1.00 \end{bmatrix} x(k) \leq \begin{bmatrix} -0.08 \\ -0.07 \\ 20.00 \end{bmatrix} \\[12pt]
\begin{bmatrix} 0.00 & 0.00 \\ 33.70 & 0.53 \end{bmatrix} x(k) + \begin{bmatrix} 0.50 \\ -0.13 \end{bmatrix} & \text{if} & \begin{bmatrix} 1.00 & -0.01 \\ 1.00 & 0.02 \\ -1.00 & -0.02 \end{bmatrix} x(k) \leq \begin{bmatrix} -0.18 \\ 0.07 \\ 0.03 \end{bmatrix} \\[12pt]
\begin{bmatrix} 0.50 \\ -1.00 \end{bmatrix} & \text{if} & \begin{bmatrix} -1.00 & -0.01 \\ 1.00 & 0.02 \\ 1.00 & -0.04 \end{bmatrix} x(k) \leq \begin{bmatrix} 0.23 \\ -0.03 \\ -0.31 \end{bmatrix} \\[12pt]
\begin{bmatrix} -0.75 & 0.03 \\ 0.00 & 0.00 \end{bmatrix} x(k) + \begin{bmatrix} 0.27 \\ -1.00 \end{bmatrix} & \text{if} & \begin{bmatrix} -1.00 & 0.04 \\ -1.00 & -0.01 \\ 1.00 & 0.01 \end{bmatrix} x(k) \leq \begin{bmatrix} 0.31 \\ 0.21 \\ -0.07 \end{bmatrix} \\[12pt]
\begin{bmatrix} -6.05 & -0.01 \\ 0.00 & 0.00 \end{bmatrix} x(k) + \begin{bmatrix} -0.09 \\ -1 \end{bmatrix} & \text{if} & \begin{bmatrix} 1.00 & 0.02 \\ -1.00 & -0.01 \\ 1.00 & 0.00 \end{bmatrix} x(k) \leq \begin{bmatrix} -0.03 \\ 0.07 \\ 0.08 \end{bmatrix} \\[12pt]
\begin{bmatrix} -0.57 & -0.01 \\ 7.75 & 0.00 \end{bmatrix} x(k) + \begin{bmatrix} -0.54 \\ -1.63 \end{bmatrix} & \text{if} & \begin{bmatrix} -1.00 & 0.00 \\ 1.00 & 0.02 \\ 0.00 & -1.00 \end{bmatrix} x(k) \leq \begin{bmatrix} -0.08 \\ -0.07 \\ 20.00 \end{bmatrix} \\[12pt]
\begin{bmatrix} 0.00 & 0.00 \\ 33.70 & 0.53 \end{bmatrix} x(k) + \begin{bmatrix} -0.50 \\ 0.13 \end{bmatrix} & \text{if} & \begin{bmatrix} -1.00 & 0.01 \\ -1.00 & -0.02 \\ 1.00 & 0.02 \end{bmatrix} x(k) \leq \begin{bmatrix} -0.18 \\ 0.07 \\ 0.03 \end{bmatrix} \\[12pt]
\begin{bmatrix} -2.84 & 0.04 \\ 34.41 & 0.52 \end{bmatrix} x(k) + \begin{bmatrix} 0 \\ 0 \end{bmatrix} & \text{if} & \begin{bmatrix} -1.00 & 0.01 \\ 1.00 & 0.02 \\ 1.00 & -0.01 \\ -1.00 & -0.02 \end{bmatrix} x(k) \leq \begin{bmatrix} 0.18 \\ 0.03 \\ 0.18 \\ 0.03 \end{bmatrix}
\end{cases}$$

Fig. 8.3 State space partition

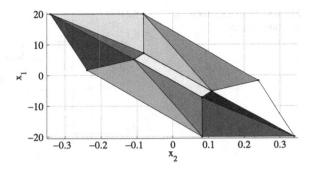

Fig. 8.4 Control inputs as piecewise affine functions of state

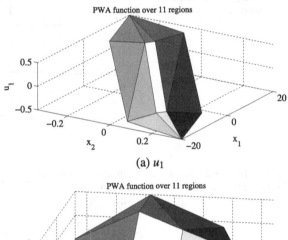

(a) u_1

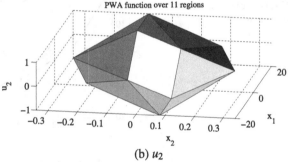

(b) u_2

Fig. 8.5 State trajectories of the closed loop system

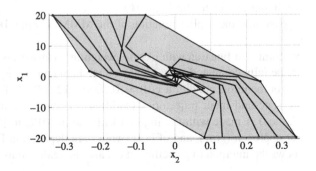

Fig. 8.6 State trajectories as functions of time

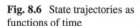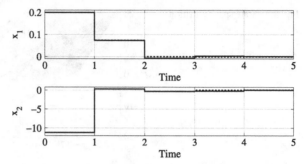

Fig. 8.7 Input trajectories as functions of time

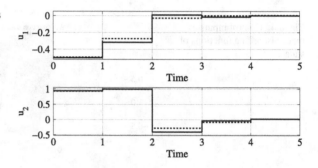

Fig. 8.8 The feasible set of [74] (white) is a subset of ours (gray)

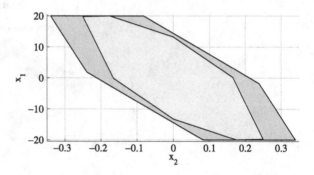

Figure 8.5 presents state trajectories of the closed loop system for different initial conditions and realizations of $\alpha(k)$.

Note that the explicit solution of the MMMPC optimization problem [21] with the
∞-norm cost function with identity weighting matrices, prediction horizon 9 could not be fully computed after 3 hours due to high complexity.

For the initial condition $x(0) = [0.2000 \; -12.0000]^T$, Fig. 8.6 and Fig. 8.7 show the state and input trajectories (solid) of the closed loop system. A comparison (dashed) is made with the implicit LMI based MPC in [74]. The feasible sets of our approach (gray), and of [74] (white) are depicted in Fig. 8.8. Finally, Fig. 8.9 shows the interpolating coefficient c^*, and the realizations of $\alpha_i(k)$, $i = 1, 2, 3, 4$.

Fig. 8.9 Interpolating
coefficient c^*, and the
realizations of $\alpha_i(k)$,
$i = 1, 2, 3, 4$ as functions of
time

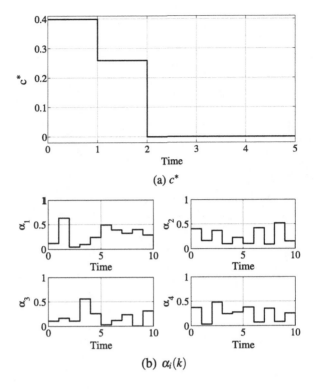

(a) c^*

(b) $\alpha_i(k)$

References

1. Ackermann, J., Bartlett, A., Kaesbauer, D., Sienel, W., Steinhauser, R.: Robust Control: Systems with Uncertain Physical Parameters. Springer, Berlin (2001)
2. Alamir, M.: Stabilization of Nonlinear System by Receding Horizon Control Scheme: A Parametrized Approach for Fast Systems (2006)
3. Alamir, M.: A framework for real-time implementation of low-dimensional parameterized NMPC. Automatica (2011)
4. Alamir, M., Bornard, G.: Stability of a truncated infinite constrained receding horizon scheme: the general discrete nonlinear case. Automatica 31(9), 1353–1356 (1995)
5. Alamo, T., Cepeda, A., Limon, D.: Improved computation of ellipsoidal invariant sets for saturated control systems. In: 44th IEEE Conference on Decision and Control, 2005 and 2005 European Control Conference, CDC-ECC'05, pp. 6216–6221. IEEE Press, New York (2005)
6. Alamo, T., Cepeda, A., Limon, D., Camacho, E.: Estimation of the domain of attraction for saturated discrete-time systems. Int. J. Inf. Syst. Sci. 37(8), 575–583 (2006)
7. Anderson, B.D.O., Moore, J.B.: Optimal Control: Linear Quadratic Methods, vol. 1. Prentice Hall, Englewood Cliffs (1990)
8. Åström, K.J., Murray, R.M.: Feedback Systems: An Introduction for Scientists and Engineers. Princeton Univ. Press, Princeton (2008)
9. Aubin, J.P.: Viability Theory. Springer, Berlin (2009)
10. Bacic, M., Cannon, M., Lee, Y.I., Kouvaritakis, B.: General interpolation in MPC and its advantages. IEEE Trans. Autom. Control 48(6), 1092–1096 (2003)
11. Bahaghighat, M.K., Kharrazi, S., Lidberg, M., Falcone, P., Schofield, B.: Predictive yaw and lateral control in long heavy vehicles combinations. In: 49th IEEE Conference on Decision and Control (CDC), pp. 6403–6408. IEEE Press, New York (2010)
12. Balas, E.: Projection with a minimal system of inequalities. Comput. Optim. Appl. 10(2), 189–193 (1998)
13. Bellman, R.: On the theory of dynamic programming. Proc. Natl. Acad. Sci. USA 38(8), 716 (1952)
14. Bellman, R.: The theory of dynamic programming. Bull. Am. Math. Soc. 60(6), 503–515 (1954)
15. Bellman, R.: Dynamic programming and Lagrange multipliers. Proc. Natl. Acad. Sci. USA 42(10), 767 (1956)
16. Bellman, R.: Dynamic programming. Science 153(3731), 34 (1966)
17. Bellman, R.E., Dreyfus, S.E.: Applied Dynamic Programming (1962)
18. Bemporad, A., Filippi, C.: Suboptimal explicit receding horizon control via approximate multiparametric quadratic programming. J. Optim. Theory Appl. 117(1), 9–38 (2003)
19. Bemporad, A., Borrelli, F., Morari, M.: Model predictive control based on linear

H.-N. Nguyen, *Constrained Control of Uncertain, Time-Varying, Discrete-Time Systems*,   189
Lecture Notes in Control and Information Sciences 451,
DOI 10.1007/978-3-319-02827-9,
© Springer International Publishing Switzerland 2014

programming—the explicit solution. IEEE Trans. Autom. Control **47**(12), 1974–1985 (2002)

20. Bemporad, A., Morari, M., Dua, V., Pistikopoulos, E.N.: The explicit linear quadratic regulator for constrained systems. Automatica **38**(1), 3–20 (2002)

21. Bemporad, A., Borrelli, F., Morari, M.: Min-max control of constrained uncertain discrete-time linear systems. IEEE Trans. Autom. Control **48**(9), 1600–1606 (2003)

22. Bitsoris, G.: Positively invariant polyhedral sets of discrete-time linear systems. Int. J. Control **47**(6), 1713–1726 (1988)

23. Blanchini, F.: Nonquadratic Lyapunov functions for robust control. Automatica **31**(3), 451–461 (1995)

24. Blanchini, F.: Set invariance in control. Automatica **35**(11), 1747–1767 (1999)

25. Blanchini, F., Miani, S.: On the transient estimate for linear systems with time-varying uncertain parameters. IEEE Trans. Circuits Syst. I, Fundam. Theory Appl. **43**(7), 592–596 (1996)

26. Blanchini, F., Miani, S.: Constrained stabilization via smooth Lyapunov functions. Syst. Control Lett. **35**(3), 155–163 (1998)

27. Blanchini, F., Miani, S.: Stabilization of LPV systems: state feedback, state estimation and duality. In: Proceedings of the 42nd IEEE Conference on Decision and Control, vol. 2, pp. 1492–1497. IEEE Press, New York (2003)

28. Blanchini, F., Miani, S.: Set-Theoretic Methods in Control. Birkhäuser, Boston (2007)

29. Blanchini, F., Miani, S.: Set-Theoretic Methods in Control. Springer, Berlin (2008)

30. Borrelli, F.: Constrained Optimal Control of Linear and Hybrid Systems. Springer, Berlin (2003)

31. Boyd, S.P., Vandenberghe, L.: Convex Optimization. Cambridge Univ. Press, Cambridge (2004)

32. Boyd, S., El Ghaoui, L., Feron, E., Balakrishnan, V.: Linear Matrix Inequalities in System and Control Theory, vol. 15. Society for Industrial Mathematics, Philadelphia (1994)

33. Bronstein, E.: Approximation of convex sets by polytopes. J. Math. Sci. **153**(6), 727–762 (2008)

34. Camacho, E.F., Bordons, C.: Model Predictive Control. Springer, Berlin (2004)

35. Campo, P.J., Morari, M.: Robust model predictive control. In: American Control Conference, pp. 1021–1026. IEEE Press, New York (1987)

36. Clarke, F.H.: Optimization and Nonsmooth Analysis vol. 5. Society for Industrial Mathematics, Philadelphia (1990)

37. Clarke, D.W., Mohtadi, C., Tuffs, P.: Generalized predictive control—part I. The basic algorithm. Automatica **23**(2), 137–148 (1987)

38. Cwikel, M., Gutman, P.O.: Convergence of an algorithm to find maximal state constraint sets for discrete-time linear dynamical systems with bounded controls and states. IEEE Trans. Autom. Control **31**(5), 457–459 (1986)

39. Dantzig, G.B.: Fourier-Motzkin elimination and its dual. Technical report, DTIC Document (1972)

40. Dreyfus, S.E.: Some types of optimal control of stochastic systems. Technical report, DTIC Document (1963)

41. Fukuda, K.: CDD/CDD+ Reference Manual. Institute for Operations Research ETH-Zentrum, ETH-Zentrum (1997)

42. Fukuda, K.: Frequently asked questions in polyhedral computation. Report, ETH, Zürich (2000) http://www.ifor.math.ethz.ch/fukuda/polyfaq/polyfaq.html

43. Gahinet, P., Nemirovskii, A., Laub, A.J., Chilali, M.: The LMI control toolbox. In: Proceedings of the 33rd IEEE Conference on Decision and Control, vol. 3, pp. 2038–2041. IEEE Press, New York (1994)

44. Genta, G.: Motor Vehicle Dynamics: Modeling and Simulation. Series on Advances in Mathematics for Applied Sciences, vol. 43. World Scientific, Singapore (1997)

45. Geyer, T., Torrisi, F.D., Morari, M.: Optimal complexity reduction of piecewise affine models based on hyperplane arrangements. In: Proceedings of the 2004 American Control Conference, vol. 2, pp. 1190–1195. IEEE Press, New York (2004)

46. Gilbert, E.G., Tan, K.T.: Linear systems with state and control constraints: the theory and application of maximal output admissible sets. IEEE Trans. Autom. Control **36**(9), 1008–1020 (1991)
47. Goodwin, G.C., Seron, M., De Dona, J.: Constrained Control and Estimation: An Optimisation Approach. Springer, Berlin (2005)
48. Grancharova, A., Johansen, T.A.: Explicit Nonlinear Model Predictive Control: Theory and Applications. Springer, Berlin (2012)
49. Grant, M., Boyd, S.: CVX: Matlab software for disciplined convex programming. Available at http://cvxr.com/cvx/
50. Grimm, G., Hatfield, J., Postlethwaite, I., Teel, A.R., Turner, M.C., Zaccarian, L.: Anti-windup for stable linear systems with input saturation: an LMI-based synthesis. IEEE Trans. Autom. Control **48**(9), 1509–1525 (2003)
51. Grünbaum, B.: Convex Polytopes, vol. 221. Springer, Berlin (2003)
52. Gutman, P.O.: A linear programming regulator applied to hydroelectric reservoir level control. Automatica **22**(5), 533–541 (1986)
53. Gutman, P.O., Cwikel, M.: Admissible sets and feedback control for discrete-time linear dynamical systems with bounded controls and states. IEEE Trans. Autom. Control **31**(4), 373–376 (1986)
54. Gutman, P.O., Cwikel, M.: An algorithm to find maximal state constraint sets for discrete-time linear dynamical systems with bounded controls and states. IEEE Trans. Autom. Control **32**(3), 251–254 (1987)
55. Hu, T., Lin, Z.: Control Systems with Actuator Saturation: Analysis and Design. Birkhäuser, Basel (2001)
56. Hu, T., Lin, Z.: Composite quadratic Lyapunov functions for constrained control systems. IEEE Trans. Autom. Control **48**(3), 440–450 (2003)
57. Hu, T., Lin, Z., Chen, B.M.: Analysis and design for discrete-time linear systems subject to actuator saturation. Syst. Control Lett. **45**(2), 97–112 (2002)
58. Hu, T., Lin, Z., Chen, B.M.: An analysis and design method for linear systems subject to actuator saturation and disturbance. Automatica **38**(2), 351–359 (2002)
59. Jiang, Z.P., Wang, Y.: Input-to-state stability for discrete-time nonlinear systems. Automatica **37**(6), 857–869 (2001)
60. Johansen, T.A., Grancharova, A.: Approximate explicit constrained linear model predictive control via orthogonal search tree. IEEE Trans. Autom. Control **48**(5), 810–815 (2003)
61. Jones, C.N., Morari, M.: The double description method for the approximation of explicit MPC control laws. In: 47th IEEE Conference on Decision and Control, CDC 2008, pp. 4724–4730. IEEE Press, New York (2008)
62. Jones, C., Morari, M.: Approximate explicit MPC using bilevel optimization. In: European Control Conference (2009)
63. Jones, C.N., Morari, M.: Polytopic approximation of explicit model predictive controllers. IEEE Trans. Autom. Control **55**(11), 2542–2553 (2010)
64. Jones, C.N., Kerrigan, E.C., Maciejowski, J.M.: Equality Set Projection: A New Algorithm for the Projection of Polytopes in Halfspace Representation. University of Cambridge, Dept. of Engineering, Cambridge (2004)
65. Jones, C., Grieder, P., Rakovic, S.: A logarithmic-time solution to the point location problem for closed-form linear MPC. In: IFAC World Congress, Prague (2005)
66. Keerthi, S., Sridharan, K.: Solution of parametrized linear inequalities by Fourier elimination and its applications. J. Optim. Theory Appl. **65**(1), 161–169 (1990)
67. Kerrigan, E.C.: Robust constraint satisfaction: invariant sets and predictive control. Department of Engineering, University of Cambridge, UK (2000)
68. Kerrigan, E.: Invariant Set Toolbox for Matlab (2003)
69. Kevrekidis, I., Aris, R., Schmidt, L.: The stirred tank forced. Chem. Eng. Sci. **41**(6), 1549–1560 (1986)
70. Khalil, H.K., Grizzle, J.: Nonlinear Systems, vol. 3. Prentice Hall, Englewood Cliffs (1992)

71. Kolmanovsky, I., Gilbert, E.G.: Theory and computation of disturbance invariant sets for discrete-time linear systems. Math. Probl. Eng. **4**(4), 317–367 (1998)
72. Kolmogorov, A.N., Fomin, S.V.: Elements of the Theory of Functions and Functional Analysis, vol. 1. Dover, New York (1999)
73. Kothare, M.V., Campo, P.J., Morari, M., Nett, C.N.: A unified framework for the study of anti-windup designs. Automatica **30**(12), 1869–1883 (1994)
74. Kothare, M.V., Balakrishnan, V., Morari, M.: Robust constrained model predictive control using linear matrix inequalities. Automatica **32**(10), 1361–1379 (1996)
75. Kurzhanskiy, A.A., Varaiya, P.: Ellipsoidal toolbox (et). In: 45th IEEE Conference on Decision and Control, pp. 1498–1503. IEEE Press, New York (2006)
76. Kvasnica, M., Grieder, P., Baotić, M., Morari, M.: Multi-parametric toolbox (MPT). Hybrid Syst., Comput. Control, 121–124 (2004)
77. Kvasnica, M., Löfberg, J., Herceg, M., Cirka, L., Fikar, M.: Low-complexity polynomial approximation of explicit MPC via linear programming. In: American Control Conference (ACC), pp. 4713–4718. IEEE Press, New York (2010)
78. Kvasnica, M., Löfberg, J., Fikar, M.: Stabilizing polynomial approximation of explicit MPC. Automatica (2011)
79. Kvasnica, M., Rauová, I., Fikar, M.: Simplification of explicit MPC feedback laws via separation functions. In: IFAC World Congress, Milano, Italy (2011)
80. Kwakernaak, H., Sivan, R.: Linear Optimal Control Systems, vol. 172. Wiley-Interscience, New York (1972)
81. Langson, W., Chryssochoos, I., Raković, S., Mayne, D.: Robust model predictive control using tubes. Automatica **40**(1), 125–133 (2004)
82. Lazar, M., de la Pena, D.M., Heemels, W., Alamo, T.: On input-to-state stability of min-max nonlinear model predictive control. Syst. Control Lett. **57**(1), 39–48 (2008)
83. Lee, E.B.: Foundations of optimal control theory. Technical report, DTIC Document (1967)
84. Lee, J., Yu, Z.: Worst-case formulations of model predictive control for systems with bounded parameters. Automatica **33**(5), 763–781 (1997)
85. Lewis, F.L., Syrmos, V.L.: Optimal Control. Wiley-Interscience, New York (1995)
86. Loechner, V., Wilde, D.K.: Parameterized polyhedra and their vertices. Int. J. Parallel Program. **25**(6), 525–549 (1997)
87. Lofberg, J.: Yalmip: a toolbox for modeling and optimization in Matlab. In: IEEE International Symposium on Computer Aided Control Systems Design, pp. 284–289. IEEE Press, New York (2004)
88. Maciejowski, J.M.: Predictive Control: With Constraints. Pearson Education, Upper Saddle River (2002)
89. Macki, J., Strauss, A.: Introduction to Optimal Control Theory. Springer, Berlin (1982)
90. Marlin, T.E.: Process Control: Designing Processes and Control Systems for Dynamic Performance. McGraw-Hill, New York (1995)
91. Marruedo, D.L., Alamo, T., Camacho, E.: Input-to-state stable MPC for constrained discrete-time nonlinear systems with bounded additive uncertainties. In: Proceedings of the 41st IEEE Conference on Decision and Control, vol. 4, pp. 4619–4624. IEEE Press, New York (2002)
92. Mayne, D.Q., Rawlings, J.B., Rao, C.V., Scokaert, P.O.M.: Constrained model predictive control: stability and optimality. Automatica **36**(6), 789–814 (2000)
93. Motzkin, T.S., Raiffa, H., Thompson, G., Thrall, R.: The Double Description Method (1953)
94. Nguyen, H.N., Olaru, S., Hovd, M.: Patchy approximate explicit model predictive control. In: International Conference on Control Automation and Systems (ICCAS), pp. 1287–1292. IEEE Press, New York (2010)
95. Olaru, S.B., Dumur, D.: A parameterized polyhedra approach for explicit constrained predictive control. In: 43rd IEEE Conference on Decision and Control, CDC, vol. 2, pp. 1580–1585. IEEE Press, New York (2004)
96. Olaru, S., Dumur, D.: On the continuity and complexity of control laws based on multiparametric linear programs. In: 45th IEEE Conference on Decision and Control, pp. 5465–5470. IEEE Press, New York (2006)

97. Olaru, S., De Doná, J.A., Seron, M., Stoican, F.: Positive invariant sets for fault tolerant multisensor control schemes. Int. J. Control **83**(12), 2622–2640 (2010)
98. Pérez, M., Albertos, P.: Self-oscillating and chaotic behaviour of a pi-controlled CSTR with control valve saturation. J. Process Control **14**(1), 51–59 (2004)
99. Pérez, M., Font, R., Montava, M.A.: Regular self-oscillating and chaotic dynamics of a continuous stirred tank reactor. Comput. Chem. Eng. **26**(6), 889–901 (2002)
100. Pistikopoulos, E.N., Georgiadis, M.C., Dua, V.: Multi-Parametric Model-Based Control: Theory and Applications, vol. 2. Vch Verlagsgesellschaft Mbh, Weinheim (2007)
101. Pistikopoulos, E.N., Georgiadis, M.C., Dua, V.: Multi-Parametric Programming: Theory, Algorithms, and Applications, vol. 1. Wiley/VCH Verlag GmbH, New York/Weinheim (2007)
102. Pluymers, B., Rossiter, J., Suykens, J., De Moor, B.: Interpolation based MPC for LPV systems using polyhedral invariant sets. In: Proceedings of the 2005 American Control Conference, pp. 810–815. IEEE Press, New York (2005)
103. Pontryagin, L.S., Gamkrelidze, R.V.: The Mathematical Theory of Optimal Processes vol. 4. CRC, Boca Raton (1986)
104. Propoi, A.: Use of linear programming methods for synthesizing sampled-data automatic systems. Autom. Remote Control **24**(7), 837–844 (1963)
105. Rakovic, S.V., Kerrigan, E.C., Kouramas, K.I., Mayne, D.Q.: Invariant approximations of the minimal robust positively invariant set. IEEE Trans. Autom. Control **50**(3), 406–410 (2005)
106. Rockafellar, R.T.: Convex Analysis, vol. 28. Princeton Univ. Press, Princeton (1997)
107. Rossiter, J.A.: Model-Based Predictive Control: A Practical Approach. CRC, Boca Raton (2003)
108. Rossiter, J., Grieder, P.: Using interpolation to improve efficiency of multiparametric predictive control. Automatica **41**(4), 637–643 (2005)
109. Rossiter, J., Kouvaritakis, B., Cannon, M.: Computationally efficient algorithms for constraint handling with guaranteed stability and near optimality. Int. J. Control **74**(17), 1678–1689 (2001)
110. Rossiter, J., Kouvaritakis, B., Bacic, M.: Interpolation based computationally efficient predictive control. Int. J. Control **77**(3), 290–301 (2004)
111. Russo, L.P., Bequette, B.W.: Effect of process design on the open-loop behavior of a jacketed exothermic CSTR. Comput. Chem. Eng. **20**(4), 417–426 (1996)
112. Scherer, C., Weiland, S.: Linear Matrix Inequalities in Control. Lecture Notes, Dutch Institute for Systems and Control, Delft, The Netherlands (2000)
113. Schmitendorf, W., Barmish, B.R.: Null controllability of linear systems with constrained controls. SIAM J. Control Optim. **18**, 327 (1980)
114. Scibilia, F., Olaru, S., Hovd, M., et al.: Approximate Explicit Linear MPC via Delaunay Tessellation (2009)
115. Seron, M.M., Goodwin, G.C., Doná, J.A.: Characterisation of receding horizon control for constrained linear systems. Asian J. Control **5**(2), 271–286 (2003)
116. Sharp, R.S., Casanova, D., Symonds, P.: Mathematical model for driver steering control, with design, tuning and performance results. Veh. Syst. Dyn. **33**(5), 289–326 (2000)
117. Skogestad, S., Postlethwaite, I.: Multivariable Feedback Control: Analysis and Design, vol. 2. Wiley, New York (2007)
118. Smith, C.A., Corripio, A.B.: Principles and Practice of Automatic Process Control, vol. 2. Wiley, New York (1985)
119. Sontag, E.D.: Algebraic approach to bounded controllability of linear systems. Int. J. Control **39**(1), 181–188 (1984)
120. Sontag, E.D., Wang, Y.: On characterizations of the input-to-state stability property. Syst. Control Lett. **24**(5), 351–359 (1995)
121. Sturm, J.F.: Using Sedumi 1.02, a Matlab toolbox for optimization over symmetric cones. Optim. Methods Softw. **11**(1–4), 625–653 (1999)
122. Summers, S., Jones, C.N., Lygeros, J., Morari, M.: A multiresolution approximation method for fast explicit model predictive control. IEEE Trans. Autom. Control **56**(11), 2530–2541 (2011)

123. Tarbouriech, S., Turner, M.: Anti-windup design: an overview of some recent advances and open problems. IET Control Theory Appl. **3**(1), 1–19 (2009)
124. Tarbouriech, S., Garcia, G., Glattfelder, A.H.: Advanced Strategies in Control Systems with Input and Output Constraints. Springer, Berlin (2007)
125. Tarbouriech, S., Garcia, G., da Silva, J.M.G., Queinnec, I.: Stability and Stabilization of Linear Systems with Saturating Actuators. Springer, Berlin (2011)
126. Taylor, C.J., Chotai, A., Young, P.C.: State space control system design based on non-minimal state-variable feedback: further generalization and unification results. Int. J. Control **73**(14), 1329–1345 (2000)
127. Tøndel, P., Johansen, T.A., Bemporad, A.: An algorithm for multi-parametric quadratic programming and explicit MPC solutions. Automatica **39**(3), 489–497 (2003)
128. Tøndel, P., Johansen, T.A., Bemporad, A.: Evaluation of piecewise affine control via binary search tree. Automatica **39**(5), 945–950 (2003)
129. Veres, S.: Geometric bounding toolbox (GBT) for Matlab. Official website: http://www.sysbrain.com (2003)
130. Wan, Z., Kothare, M.V.: An efficient off-line formulation of robust model predictive control using linear matrix inequalities. Automatica **39**(5), 837–846 (2003)
131. Wang, Y., Boyd, S.: Fast model predictive control using online optimization. IEEE Trans. Control Syst. Technol. **18**(2), 267–278 (2010)
132. Zaccarian, L., Teel, A.R.: A common framework for anti-windup, bumpless transfer and reliable designs. Automatica **38**(10), 1735–1744 (2002)
133. Ziegler, G.M.: Lectures on Polytopes. Springer, Berlin (1995)

Index

H.-N. Nguyen, *Constrained Control of Uncertain, Time-Varying, Discrete-Time Systems*, 195
Lecture Notes in Control and Information Sciences 451,
DOI 10.1007/978-3-319-02827-9,
© Springer International Publishing Switzerland 2014